SpringerBriefs in Computer Science

W0230288

SpringerBriefs present concise summaries of cutting-edge research and practical applications across a wide spectrum of fields. Featuring compact volumes of 50 to 125 pages, the series covers a range of content from professional to academic.

Typical topics might include:

- A timely report of state-of-the art analytical techniques
- A bridge between new research results, as published in journal articles, and a contextual literature review
- A snapshot of a hot or emerging topic
- An in-depth case study or clinical example
- A presentation of core concepts that students must understand in order to make independent contributions

Briefs allow authors to present their ideas and readers to absorb them with minimal time investment. Briefs will be published as part of Springer's eBook collection, with millions of users worldwide. In addition, Briefs will be available for individual print and electronic purchase. Briefs are characterized by fast, global electronic dissemination, standard publishing contracts, easy-to-use manuscript preparation and formatting guidelines, and expedited production schedules. We aim for publication 8–12 weeks after acceptance. Both solicited and unsolicited manuscripts are considered for publication in this series.

More information about this series at http://www.springer.com/series/10028

Yuanwei Liu • Zhijin Qin • Zhiguo Ding

Non-Orthogonal Multiple Access for Massive Connectivity

 Springer

Yuanwei Liu
London, UK

Zhijin Qin
London, UK

Zhiguo Ding
Manchester, UK

ISSN 2191-5768   ISSN 2191-5776   (electronic)
SpringerBriefs in Computer Science
ISBN 978-3-030-30974-9   ISBN 978-3-030-30975-6   (eBook)
https://doi.org/10.1007/978-3-030-30975-6

This Springer imprint is published by the registered company Springer Nature Switzerland AG.
The registered company address is: Gewerbestrasse 11, 6330 Cham, Switzerland

How to support massive number of devices for future wireless networks?

Foreword

Non-orthogonal multiple access (NOMA) holds great promise for meeting the phenomenal increase in demand for both wireless user access and capacity, fueled by the Internet of Things (IoT). Unlike orthogonal multiple-access schemes, in NOMA, multiple users can share the same time or frequency resource while being assigned different codes or power levels and separated at the receiver using successive interference cancelation techniques. *Non-Orthogonal Multiple Access for Massive Connectivity* is a much-needed reference on this critically important technology for 5G and beyond networks. This first-of-its-kind book from the experts on this subject presents a comprehensive framework for the design and analysis of power-domain NOMA divided into three main parts that address the key issues of compatibility, sustainability, and security. In the compatibility part, the authors masterfully demonstrate the seamless integration of NOMA with other key wireless technologies such as multi-input multi-output (MIMO) and its benefits when applied to cognitive radio networks and heterogeneous networks, where the whole becomes greater than the sum of its parts. In the sustainability part, the authors skillfully show how NOMA can be efficiently integrated with cooperative communication and simultaneous wireless information and power transfer (SWIPT) protocols to extend network reliability and lifetime. In the security part, the authors brilliantly analyze the physical layer security performance of NOMA networks and quantify the secrecy gains possible with the aid of artificial noise signals. Finally, the authors explore two exciting topics with lots of intriguing unanswered questions for future research; namely, the application of NOMA to unmanned aerial vehicle (UAV) networks and the exploitation of machine learning tools to further enhance the performance of NOMA-based wireless networks. This well-written book provides an in-depth treatment of the subject and strikes an excellent balance between theory and practice. It will serve as a valuable reference on NOMA for researchers and practicing engineers for years to come.

Richardson, TX, USA   Naofal Al-Dhahir
July 2019

Preface

In this book, we discuss NOMA and the various issues in NOMA networks, including capability, sustainability, and security. This book starts from the basics and key techniques of NOMA. Subsequently, we identify three critical issues in NOMA networks, including compatibility, sustainability, and security. Particularly, we first demonstrate the applications of NOMA in different networks including MIMO-NOMA, NOMA in heterogeneous networks, and NOMA in cognitive radio networks to show the compatibility of NOMA with various networks. Then, the wireless-powered NOMA networks are presented to address the sustainability issues in NOMA networks to extend the network reliability and lifetime. The security-enhanced NOMA networks are discussed for single antenna case and multiple antenna case, respectively. Finally, the most recent developments on artificial intelligence (AI)-enabled NOMA networks are discussed, and the research challenges on NOMA to support massive number of devices are identified. We believe this book will provide readers a clear picture on the performance and benefits of adopting NOMA for the next generation of wireless communication systems to support massive connectivity.

Acknowledgement

I would like to express my sincere gratitude to all the colleagues who contributed to the work and projects that led to this book.

I would also like to particularly thank our editor as well as all the editorial staffs from Springer for producing this book.

London, UK   Yuanwei Liu
April 2019

Contents

Acronyms

AF	Amplify-and-Forward
AI	Artificial Intelligence
AWGN	Additive White Gaussian Noise
BB	Beamforming-Based
BC	Broadcast Channel
BF	Beamforming
BS	Base Station
CB	Cluster-Based
CDF	Cumulative Distribution Function
CDMA	Code Division Multiple Access
CF	Compress-and-Forward
CIRs	Channel Impulse Responses
CoMP	Coordinated Multipoint
CR	Cognitive Radio
CSI	Channel State Information
CSIT	Channel State Information at the Transmitter
D2D	Device-to-Device
DF	Decode-and-Forward
DL	Downlink
FDMA	Frequency Division Multiple Access
IDMA	Interleave Division Multiple Access
IMD	Iterative Multi-user Detection
IoT	Internet of Things
LDPC	Low-Density Parity-Check
LDS	Low-Density Signature
LMMSE	Linear Minimum Mean Square Error
LPMA	Lattice Partition Multiple Access
LTE	Long Term Evolution
M2M	Machine-to-Machine
MA	Multiple Access
MAC	Medium Access Control

ML	Machine Learning
MNV	Wireless Network Visualization
MPA	Message Passing Algorithms
MUSA	Multi-User Shared Access
MUST	Multi-User Superposition Transmission
NP	Non-deterministic Polynomial time
NOMA	Non-Orthogonal Multiple Access
OFDM	Orthogonal Frequency Division Multiplexing
OFDMA	Orthogonal Frequency Division Multiple Access
OMA	Orthogonal Multiple Access
P2P	Peer-to-Peer
PA	Power Allocation
PDMA	Pattern Division Multiple Access
PLS	Physical Layer Security
PR	Primary Receiver
PT	Primary Transmitter
PU	Primary User
QoS	Quality of Service
RB	Resource Block
RBC	Relaying Broadcast Channel
RF	Radio Frequency
SA	Signal Alignment
SC	Superposition Coding
SCMA	Sparse Code Multiple Access
SDM	Space Division Multiplexing
SDMA	Space Division Multiple Access
SDN	Software-Defined Network
SD-NOMA	Software-Defined NOMA
SDR	Software-Defined Radio
SIC	Successive Interference Cancelation
SISO	Single-Input Single-Output
SNR	Signal-Noise Ratio
SR	Secondary Receiver
ST	Secondary Transmitter
SU	Secondary User
SWIPT	Simultaneous Wireless Information and Power Transfer
TCMA	Trellis Coded Multiple Access
TDMA	Time Division Multiple Access
UL	Uplink
WPT	Wireless Power Transfer
ZF	Zero-forcing

Part I
Background

Chapter 1
Introduction

This chapter introduces the background of using non-orthogonal multiple access (NOMA), including overview of NOMA in the fifth generation (5G) communication systems and the beyond as well as the related standards.

1.1 Background

Driven by the rapid escalation of the wireless capacity requirements imposed by advanced multimedia applications (e.g., ultra-high-definition video, virtual reality, etc.), as well as the dramatically increasing demand for user access required by Internet of Things (IoT) devices (Qin et al. 2019), the 5G and beyond networks face challenges in terms of supporting large-scale heterogeneous data traffic. The often-quoted albeit potentially unrealistic expectations include 1000 times higher system capacity, 10 times higher system throughput, and 10 times higher energy efficiency per service than those of the fourth generation (4G) networks. Several key directions have been identified by researchers. Sophisticated multiple-access (MA) techniques have been regarded as one of the most fundamental enablers, which have significantly evolved over the consecutive generations in wireless networks (Liu et al. 2017; Cai et al. 2018).

Looking back to the development of the MA formats, in the first generation (1G), FDMA was combined with an analog frequency modulation-based technology, although digital control channel signaling was used. In the second generation (2G) GSM communications time division multiple access (TDMA) was used (Steele and Hanzo 1999). Then code division multiple access (CDMA), originally proposed by Qualcomm (Gilhousen et al. 1991), became the dominant MA in the 3G networks. In an effort to overcome the inherent limitation of CDMA—namely that the chip rate has to be much higher than the information data rate—orthogonal frequency division multiple access (OFDMA) was adopted for the 4G networks. Based on whether the same time or frequency resource can be occupied by more than one user, the existing MA techniques may be categorized into OMA and NOMA techniques. Among the above-mentioned MA techniques, FDMA, TDMA, and OFDMA allow only a single user to be served within the same time/frequency resource block (RB), which belong to the OMA approach. By contrast, CDMA allows multiple users to be supported by the same RB with the aid of applying different unique, user-specific spreading sequences for distinguishing them.

Specifically, NOMA techniques can be primarily classified into a pair of categories, namely, *code-domain* NOMA (Yuan et al. 2018) and *power-domain* NOMA. The representatives *code-domain NOMA* techniques include trellis coded multiple access (TCMA), interleave division multiple access (IDMA), low-density signature (LDS) sequence-based CDMA. These solutions are complemented by the more recently proposed multi-user shared access (MUSA) technique, pattern division multiple access (PDMA), and sparse code multiple access (SCMA).

The *power-domain NOMA*, which has been recently proposed to 3GPP LTE (Ding et al. 2017), exhibits a superior capacity region compared to OMA. The key idea of power-domain NOMA is to ensure that multiple users can be served within a given time/frequency RB, with the aid of superposition coding (SC) techniques at the transmitter and successive interference cancelation (SIC) at the receiver, which is fundamentally different from the classic OMA techniques of FDMA/TDMA/OFDMA as well as from the code-domain NOMA techniques. The motivation behind this approach lies in the fact that NOMA is capable of exploiting the available resources more efficiently by opportunistically capitalizing on the users' specific channel conditions and it is capable of serving multiple users at different QoS requirements in the same RB. It has also been pointed out that NOMA has the potential to be integrated with existing MA paradigms, since it exploits the new dimension of the power domain.

1.2   Standardization of NOMA

NOMA has recently been included into LTE-A, terms MUST (Lee et al. 2016). Specifically, at the 3GPP meeting in May 2015, it was decided to include MUST into LTE Advanced. Afterwards, at the 3GPP meeting in August 2015, 15 different forms of MUST have been proposed by Huawei, Qualcomm, NTT DOCOMO, etc. Finally, at the 3GPP meeting in December 2015, NOMA has been included into LTE Release 13 (3rd Generation Partnership Project (3GPP) 2015). It is worth noting that the MUST technique may be made compatible with the existing LTE structure. In other words, NOMA allows two users to be served at the same OFDM subcarrier without changing the current structure. Various non-orthogonal transmission schemes have been proposed for the MUST in 3rd Generation Partnership Project (3GPP) (2015) and Lee et al. (2016), which can be generally classified into three categories according to 3rd Generation Partnership Project (3GPP) R1-154999 (2015), namely (1) superposition transmission with an adaptive power ratio on each component constellation and non-Gray-mapped composite constellation; (2) superposition transmission with an adaptive power ratio on component constellations and Gray-mapped composite constellation; (3) superposition transmission with a label-bit assignment on composite constellation and Gray-mapped composite constellation.

In addition to LTE-A, NOMA had also been included in the forthcoming digital TV standard, by the Advanced Television Systems Committee (ATSC) 3.0 (Zhang et al. 2016), termed as layered division multiplexing (LDM) for providing significant improvements in terms of service reliability, system flexibility, and spectrum efficiency. This standard will generate significant impact on digital TV industry. Moreover, in the white paper of NTT DOCOMO, NOMA has been identified as a key technique for 5G. The system-level performance of NOMA has also been demonstrated by NTT DOCOMO (Benjebbour et al. 2013). The key challenges of implementing NOMA in industry is the decoding complexity increases at receivers as the number of users increase. Another potential challenge is that the security and privacy of far users should be protected at near user side due to the characteristic of SIC, which may depend on key generations from upper layer. Although NOMA is reviewed as a promising candidate for future networks, there are various forms of NOMA. The standardization process for NOMA is still ongoing.

References

3rd Generation Partnership Project (3GPP) (2015). Study on downlink multiuser superposition transmission for LTE.

3rd Generation Partnership Project (3GPP) R1-154999 (2015). TP for classification of MUST schemes.

Benjebbour, A., Li, A., Saito, Y., Kishiyama, Y., Harada, A., & Nakamura, T. (2013). System-level performance of downlink NOMA for future LTE enhancements. In *Proceedings IEEE Global Communications Conference (GLOBECOM) Workshops* (pp. 66–70).

Cai, Y., Qin, Z., Cui, F., Li, G. Y., & McCann, J. A. (2018). Modulation and multiple access for 5G networks. *IEEE Communications Surveys & Tutorials, 20*, 629–646.

Ding, Z., Liu, Y., Choi, J., Sun, Q., Elkashlan, M., Chih-Lin., I., et al. (2017). Application of non-orthogonal multiple access in LTE and 5G networks. *IEEE Communications Magazine, 55*, 185–191.

Gilhousen, K. S., Jacobs, I. M., Padovani, R., Viterbi, A. J., Weaver Jr, L. A., & Wheatley III, C. E. (1991). On the capacity of a cellular CDMA system. *IEEE Transactions on Vehicular Technology, 40*, 303–312

Lee, H., Kim, S., & Lim, J. H. (2016). Multiuser superposition transmission (MUST) for LTE-A systems. In *IEEE Proceedings of International Conference on Communications (ICC)* (pp. 1–6).

Liu, Y., Qin, Z., Elkashlan, M., Ding, Z., Nallanathan, A., & Hanzo, L. (2017). Nonorthogonal multiple access for 5G and beyond. *Proceedings of the IEEE, 105*, 2347–2381.

Qin, Z., Li, F. Y., Li, G. Y., McCann, J. A., & Ni, Q. (2019). Low-power wide-area networks for sustainable IoT. *IEEE Wireless Communications, 26*, 1–6.

Steele, R., & Hanzo, L. (1999). *Mobile radio communications: Second and third generation cellular and WATM systems: 2nd.* Piscataway, Hoboken: IEEE Press-John Wiley.

Yuan, L., Pan, J., Yang, N., Ding, Z., & Yuan, J. (2018). Successive interference cancellation for LDPC coded nonorthogonal multiple access systems. *IEEE Transactions on Vehicular Technology, 67*, 5460–5464.

Zhang, L., Li, W., Wu, Y., Wang, X., Park, S. I., Kim, H. M., et al. (2016). Layered-division-multiplexing: Theory and practice. *IEEE Transactions on Broadcasting, 62*(1), 216–232.

Chapter 2
What Is NOMA?

This chapter provides an overview of the background knowledge of NOMA from an information theoretic perspective.

2.1 NOMA Basis

2.1.1 Investigating NOMA from an Information Theoretic Perspective

Having considered the potential benefits of NOMA, it is important to investigate its performance gain also from an information theoretic perspective. In fact, the concept of NOMA may also be interpreted as a special case of SC in the downlink broadcast channel (BC). More particularly, by using SC, the capacity region of a realistic imperfect discrete memoryless BC was established by Cover (1972). As an extension of Cover (1972), Bergmans found the Gaussian BC capacity region of single-antenna scenarios (Bergmans 1974). Inspired by Cover (1972) and Bergmans (1974), several researchers began to explore the potential performance gain from an information theoretic perspective. Xu et al. (2015) developed a new evaluation criterion for quantifying the performance gain of NOMA over OMA. More specifically, considering a simple two-user single-antenna scenario in conjunction with the Gaussian BC, the comparison of TDMA and NOMA in terms of their capacity region was provided in Xu et al. (2015). The analytical results showed that NOMA is capable of outperforming TDMA in terms of both the individual user-rates and the sum rate. Shieh and Huang (2016) focused their attentions on examining the capacity region of downlink NOMA by systematically designing practical schemes and investigated the gains of NOMA over OMA by designing practical encoders and decoders. Furthermore, by proposing to use NOMA for relaying broadcast channels (RBC) for the sake of achieving a performance enhancement, So and

Y. Liu et al., *Non-Orthogonal Multiple Access for Massive Connectivity*, SpringerBriefs in Computer Science, https://doi.org/10.1007/978-3-030-30975-6_2

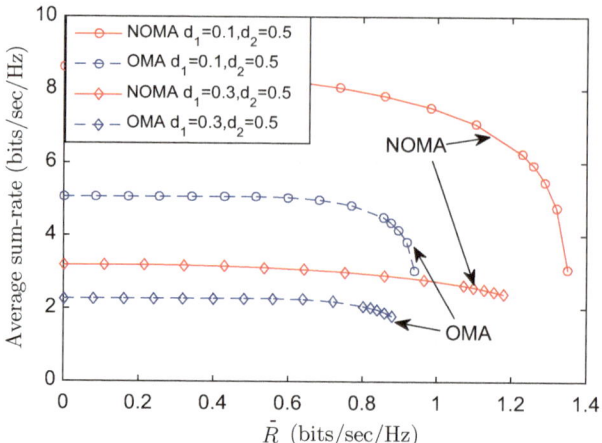

Fig. 2.1 Illustration of the sum-rate versus the minimum rate constraints with full CSIT knowledge for different user distances. The full parameter settings can be found in Xing et al. (2017)

Sung (2016) examined the achievable capacity region of the RBC upon invoking both decode-and-forward (DF) relaying and compress-and-forward (CF) relaying with/without dirty-paper coding (DPC). Regarding the family of MIMO-NOMA systems, in Liu et al. (2016) the achievable capacity region of multiuser MIMO systems is investigated upon invoking iterative linear minimum mean square error (LMMSE) detection.

In contrast to the above research contributions considering NOMA in additive white Gaussian noise (AWGN) channels, Xing et al. (2017) investigated the performance of a two-user case of downlink NOMA in fading channels to exploit the time-varying nature on multi-user channels. As shown in Fig. 2.1, NOMA achieves a superior performance over OMA for different distance settings, where the average sum-rate was maximized subject to a minimum average individual rate constraint.

2.1.2 Downlink NOMA Transmission

Downlink NOMA transmission employs the SC technique at the BS for sending a combination of the signals and the SIC technique may be invoked by the users for interference cancelation, as shown in Fig. 2.2b. Numerous valuable contributions have investigated the performance of NOMA in terms of downlink transmission (Saito et al. 2013a,b; Benjebbour et al. 2013; Ding et al. 2014; Timotheou and Krikidis 2015; Huang and Yang 2019; Sun et al. 2018; Yu et al. 2018). In Saito et al. (2013b), a two-user NOMA downlink transmission relying on SIC receivers was proposed. Upon considering a range of further practical conditions in terms of the key link-adaptation functionalities of the LTE, the system-level performance

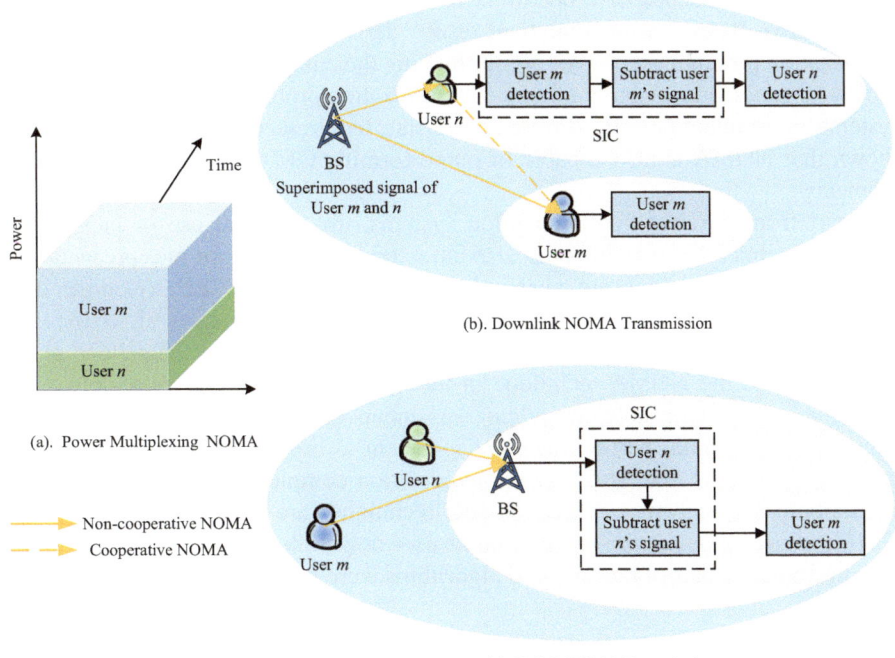

Fig. 2.2 Illustration of NOMA transmission. (**a**) Power multiplexing NOMA. (**b**) Downlink NOMA transmission. (**c**) Uplink NOMA transmission

was evaluated in Saito et al. (2013a) and Benjebbour et al. (2013). A more general NOMA transmission scheme was proposed in Ding et al. (2014), which considered a BS communicating with M randomly deployed users. It was demonstrated that NOMA is capable of achieving superior performance compared to OMA in terms of both its outage probability and its ergodic rate. In Timotheou and Krikidis (2015), the fairness issues were addressed by adopting appropriate PA coefficients for the M users in a general NOMA downlink transmission scenario.

Motivated by reducing the signaling overhead required for CSI estimation, some researchers embarked on investigating the performance of downlink NOMA transmissions using partial CSI at the transmitter (Yang et al. 2016; Shi et al. 2016; Cui et al. 2016). More explicitly, Yang et al. (2016) studied the outage probability of NOMA by assuming either imperfect CSI or second order statistics-based CSI, respectively. By assuming the knowledge of statistical CSI, Shi et al. (2016) investigated the outage performance of NOMA by jointly considering both the decoding order selection and the PA of the users. By assuming that only the average CSI was obtained at the BS, Cui et al. (2016) studied both the optimal decoding order and the optimal PA of the users in downlink NOMA systems. Both the transmit power of the BS and rate-fairness of users were optimized. By considering only a single-bit feedback of the CSI from each user to the BS, the outage performance of a downlink transmission scenario was studied by Xu et al. (2016). Based on the analytical results derived, the associated dynamic PA optimization problem was solved by minimizing the outage probability. Zhang et al. (2017) investigated an energy efficiency optimization problem in downlink NOMA systems in conjunction with different data rate requirements of the users. It was shown that NOMA is also capable of outperforming OMA in terms of its energy efficiency.

Apart from wireless communication systems, the potential use of NOMA in other communication scenarios has also attracted interests. A pair of representative communication scenarios are visible light communication (VLC) (Komine and Nakagawa 2004; Marshoud et al. 2016; Yin et al. 2016; Zhang et al. 2016) and quantum-aided communication (Hanzo et al. 2012). To elaborate, Marshoud et al. (2016) applied the NOMA technique in the context of VLC downlink networks. By doing so, the achievable throughput was enhanced. It is worth pointing out that although the potential performance enhancement is brought by invoking NOMA into VLC networks, the hardware implementation complexity is also increased at transmitters and receivers as SC and SIC techniques are adopted. Botsinis et al. (2016) considered quantum-assisted multi-user downlink transmissions in NOMA systems, where a pair of bio-inspired algorithms were proposed.

2.1.3   Uplink NOMA Transmission

In uplink NOMA transmission, multiple users transmit their own uplink signals to the BS in the same RB, as shown in Fig. 2.2c. The BS detects all the messages of the users with the aid of SIC. Note that there are several key differences between uplink NOMA and downlink NOMA, which are listed as follows:

- **Transmit Power:** In contrast to downlink NOMA, the transmit power of the users in uplink NOMA does not have to be different, and it depends on the channel conditions of each user. If the users' channel conditions are significantly different, their received SINR can be rather different at the BS, regardless of their transmit power.
- **SIC Operations:** The SIC operations and interference experienced by the users in the uplink NOMA and downlink NOMA are also rather different. More specifically, as shown in Fig. 2.2b for downlink NOMA, the signal of User n is decontaminated from the interference imposed by User m, which is achieved by first detecting the stronger signal of User m, remodulating it, and then subtracting it from the composite signal. It means that SIC operation is carried out on strong user in downlink for canceling the weak user's interference. By contrast, in uplink NOMA, SIC is carried out at the BS to detect strong User n first by treating User m as interference, as shown in Fig. 2.2c. Then it remodulates the recovered signal and subtracts the interference imposed by User n to detect User m.

- **Performance Gain:** The performance gain of NOMA over OMA is different for downlink and uplink. Figure 1 of Shin et al. (2016) illustrates the capacity region of NOMA and OMA both for downlink and uplink. The capacity region of NOMA is outside OMA, which means that the use of NOMA in downlink has superior performance in terms of throughput. While in uplink systems, NOMA mainly has the advantages in terms of fairness, especially compared to that OMA with power control.

References

Benjebbour, A., Li, A., Saito, Y., Kishiyama, Y., Harada, A., & Nakamura, T. (2013). System-level performance of downlink NOMA for future LTE enhancements. In *Proceedings of IEEE Global Communications Conference (GLOBECOM) Workshops* (pp. 66–70).

Bergmans, P. (1974). A simple converse for broadcast channels with additive white Gaussian noise (corresp.). *IEEE Transactions on Information Theory, 20*, 279–280.

Botsinis, P., Alanis, D., Babar, Z., Nguyen, H., Chandra, D., Ng, S. X., et al. (2016). Quantum-aided multi-user transmission in non-orthogonal multiple access systems. *IEEE Access, 4*, 7402–7424.

Cover, T. M. (1972). Broadcast channels. *IEEE Transactions on Information Theory, 18*, 2–14.

Cui, J., Ding, Z., & Fan, P. (2016). A novel power allocation scheme under outage constraints in NOMA systems. *IEEE Signal Processing Letters, 23*, 1226–1230.

Ding, Z., Yang, Z., Fan, P., & Poor, H. V. (2014). On the performance of non-orthogonal multiple access in 5G systems with randomly deployed users. *IEEE Signal Processing Letters, 21*, 1501–1505.

Hanzo, L., Haas, H., Imre, S., O'Brien, D., Rupp, M., & Gyongyosi, L. (2012). Wireless myths, realities, and futures: From 3G/4G to optical and quantum wireless. *IEEE Proceedings, 100*, 1853–1888.

Huang, X., & Yang, N. (2019). On the block error performance of short-packet non-orthogonal multiple access systems. In *IEEE Proceedings of International Communication Conference (ICC)* (pp. 1–7).

Komine, T., & Nakagawa, M. (2004). Fundamental analysis for visible-light communication system using LED lights. *IEEE Transactions on Consumer Electronics, 50*, 100–107.

Liu, L., Yuen, C., Guan, Y. L., & Li, Y. (2016). Capacity-achieving iterative LMMSE detection for MIMO-NOMA systems. *IEEE Transactions on Signal Processing, 67*(7), 1758–1773. 1 April 2019.

Marshoud, H., Kapinas, V. M., Karagiannidis, G. K., & Muhaidat, S. (2016). Non-orthogonal multiple access for visible light communications. *IEEE Photonics Technology Letters, 28*, 51–54.

Saito, Y., Benjebbour, A., Kishiyama, Y., & Nakamura, T. (2013a). System-level performance evaluation of downlink non-orthogonal multiple access (NOMA). In *Proceedings of IEEE Annual Symposium on Personal, Indoor and Mobile Radio Communications (PIMRC), London.*

Saito, Y., Kishiyama, Y., Benjebbour, A., Nakamura, T., Li, A., & Higuchi, K. (2013b). Non-orthogonal multiple access (NOMA) for cellular future radio access. In *IEEE Proceedings of Vehicular Technology Conference (VTC), Dresden* (pp. 1–5).

Shi, S., Yang, L., & Zhu, H. (2016). Outage balancing in downlink non-orthogonal multiple access with statistical channel state information. *IEEE Transactions on Wireless Communications, 15*, 4718–4731.

Shieh, S. L., & Huang, Y. C. (2016). A simple scheme for realizing the promised gains of downlink non-orthogonal multiple access. *IEEE Transactions on Communications, 64*, 1624–1635.

Shin, W., Vaezi, M., Lee, B., Love, D. J., Lee, J., & Poor, H. V. (2016). Non-orthogonal multiple access in multi-cell networks: Theory, performance, and practical challenges. arXiv preprint arXiv:1611.01607.

So, J., & Sung, Y. (2016). Improving non-orthogonal multiple access by forming relaying broadcast channels. *IEEE Communications Letters, 20*, 1816–1819.

Sun, X., Yan, S., Yang, N., Ding, Z., Shen, C., & Zhong, Z. (2018). Short-packet downlink transmission with non-orthogonal multiple access. *IEEE Transactions on Wireless Communications, 17*, 4550–4564.

Timotheou, S., & Krikidis, I. (2015). Fairness for non-orthogonal multiple access in 5G systems. *IEEE Signal Processing Letters, 22*, 1647–1651.

Xing, H., Liu, Y., Nallanathan, A., Ding, Z., & Poor, H. V. (2018). Optimal Throughput Fairness Tradeoffs for Downlink Non-Orthogonal Multiple Access Over Fading Channels. In *IEEE Transactions on Wireless Communications, 17*(6), vol. 17, no. 6, pp. 3556–3571, June 2018.

Xu, P., Ding, Z., Dai, X., & Poor, H. V. (2015). A new evaluation criterion for non-orthogonal multiple access in 5G software defined networks. *IEEE Access, 3*, 1633–1639.

Xu, P., Yuan, Y., Ding, Z., Dai, X., & Schober, R. (2016). On the outage performance of non-orthogonal multiple access with one-bit feedback. *IEEE Transactions on Wireless Communications, 15*, 6716–6730.

Yang, Z., Ding, Z., Fan, P., & Karagiannidis, G. K. (2016). On the performance of non-orthogonal multiple access systems with partial channel information. *IEEE Transactions on Communications, 64*, 654–667.

Yin, L., Popoola, W. O., Wu, X., & Haas, H. (2016). Performance evaluation of non-orthogonal multiple access in visible light communication. *IEEE Transactions on Communications, 64*, 5162–5175.

Yu, H., Fei, Z., Yang, N., & Ye, N. (2018). Optimal design of resource element mapping for sparse spreading non-orthogonal multiple access. *IEEE Wireless Communications Letters, 7*, 744–747.

Zhang, X., Gao, Q., Gong, C., & Xu, Z. (2016). Interference management and power allocation for NOMA visible light communications network. arXiv preprint. arXiv:1610.07327.

Zhang, Y., Wang, H.-M., Zheng, T.-X., & Yang, Q. (2017). Energy-efficient transmission design in non-orthogonal multiple access. *IEEE Transactions on Vehicular Technology, 66*, 2852–2857.

Part II
NOMA in Future Wireless Networks

Chapter 3
Compatibility in NOMA

In this chapter, the compatibility of NOMA will be introduced by discussing the applications of NOMA to various techniques, such as heterogeneous networks (HetNets), cognitive radio networks (CRNs), and multiple-input multiple-output (MIMO). Particularly, the average performance of NOMA enabled HetNets will be provided as an example.

3.1 NOMA in Heterogeneous Networks

HetNets and massive multiple-input multiple-output (MIMO), as two of the "big three" technologies, laid the fundamental structure for future network designs. The massive MIMO regime enables to equip tens of hundreds/thousands antennas at a BS, and hence is capable of offering an unprecedented level of freedom to serve multiple mobile users. The core idea of HetNets is to establish closer BS-user link by densely overlaying small cells. By doing so, the promising benefits such as lower power consumption, higher throughput, and enhanced spatial reuse of spectrum can be experienced. Aiming to fully take advantages of both massive MIMO and HetNets, several research contributions have been made (Adhikary et al. 2015; Ye et al. 2015; Liu et al. 2016b). In Adhikary et al. (2015), the interference coordination issue of massive MIMO enabled HetNets was addressed by utilizing the spatial blanking of macro cells. In Ye et al. (2015), the authors investigated a joint user association and interference management optimization problem in massive MIMO HetNets. By applying stochastic geometry model, the spectrum efficiency of uplink massive MIMO-aided HetNets was evaluated in Liu et al. (2016b).

Among the recent research contributions towards 5G and the beyond, NOMA-based HetNets has not been well investigated yet and is still in its infancy. We believe that the novel structure design in this work—by introducing NOMA-based small cells in massive MIMO enabled HetNets—can be a new highly rewarding candidate, which will contribute to the design of a more promising future wireless networks due to the following key advantages:

- High spectrum efficiency: NOMA improves the spectrum efficiency with multiplexing users in power domain and invoking successive interference cancelation (SIC) technique for canceling interference. In NOMA-based HetNets, with employing higher BS densities, BSs are capable of accessing the served users closer, which can increase the signal-to-interference-plus-noise ratio (SINR) by intelligently tracking the multi-category interference, such as inter/intra-tier interference and intra-BS interference.
- High compatibility and low complexity: NOMA is regarded as a promising "add-on" technology for the existing multiple-access systems due to the gradually mature of superposition coding (SC) and SIC technologies, and will not bring much implementation complexity. Additionally, with applying NOMA in the single-antenna-based small cells, the complex precoding/cluster design for MIMO-NOMA systems can be avoided.
- Fairness/throughput tradeoff: NOMA is capable of dealing with the fairness issue by allocating more power to weak users, which is of great significance for HetNets when investigating efficient resource allocation in the sophisticated multi-tier networks.

NOMA-based HetNets will not bring much implementation complexity or modification for the existing networks. Additionally, with applying NOMA in the single-antenna-based small cells, the complex precoding/cluster design for the multi-antenna NOMA can be avoided.

3.1.1 Network Model

3.1.1.1 Network Description

Motivated by the aforementioned potential benefits, we propose a novel hybrid HetNets framework with NOMA-based small cells and massive MIMO-aided macro cells to further enhance the performance of existing HetNets design. In this framework, we consider a downlink K-tier HetNets, where the first tier represents the macro cells and the other tiers represent the small cells such as pico cells and femto cells. The positions of macro BSs and all the k-th tier ($k \in \{1, \ldots, K\}$) BSs are modeled as homogeneous poisson point processes (HPPPs) Φ_k and with density λ_k, respectively. α_k is the path loss exponent of the k-th tier cells. All channels are assumed to undergo quasi-static Rayleigh fading, where the channel coefficients are constant for each transmission block but independent between different blocks.

Motivated by the fact that it is common to overlay a high-power macro cell with successively denser and lower power small cell, we consider to apply massive

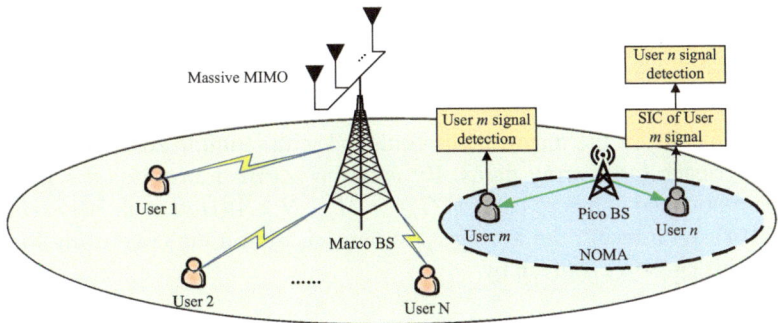

Fig. 3.1 Illustration of NOMA and massive MIMO-based hybrid HetNets

MIMO technologies to macro cells and NOMA transmission to small cells in this work. As shown in Fig. 3.1, macro BSs are considered to be equipped with M antennas, each macro BS transmit signals to N users over the same resource block (e.g., time/frequency/code).[1] We assume $M \gg N > 1$ and the linear ZFBF technique is applied at each macro BS with assigning equal power to N data streams. In small cells, each small cell BS is considered to be equipped with single antenna. In other words, in this scenario, macro cells are OMA based and small cells are NOMA based. All users are considered to be equipped with single antenna each as well. We consider to adopt user pairing in each tier of small cells to implement NOMA for lowering the system complexity (Liu et al. 2016c; Qin et al. 2018b).

3.1.1.2 NOMA and Massive MIMO-Based User Association

In this work, a user is allowed to access any tier BS, which provides the best coverage. We consider the flexible user association which is based on the maximum average received power of each tier.

Different from the convectional user association in OMA, NOMA exploits the power sparsity for multiple access by allocating different powers to different users. Due to the random spatial topology of the stochastic geometry model, the space information of users are not predetermined (Qin et al. 2019). The user association policy for the NOMA enhanced small cells assumes that near user is chosen as the typical one first. As such, at the i-th tier small cell, the averaged power received at users connecting to the i-th tier BS j (where $j \in \Phi_i$) is given by

$$P_{r,i} = a_{n,i} P_i L\left(d_{j,i}\right) B_i, \tag{3.1}$$

---

[1] The aim is to avoid sophisticated MIMO-NOMA design in macro cells.

where P_i is the transmit power of a i-th tier BS, $a_{n,i}$ is the power sharing coefficient for the near user, $L\left(d_{j,i}\right) = \eta d_{j,i}^{-\alpha_i}$ is large-scale path loss, $d_{j,i}$ is the distance between the user and a i-th tier BS, and α_i is the path loss exponent of the i-th tier small cell.

In macro cells, as the macro BS is equipped with multiple antennas, macro cell users experience large array gains. By adopting ZFBF transmission scheme, the array gain obtained at macro users is $G_M = M - N + 1$ (Huh et al. 2012; Hosseini et al. 2014). As a result, the average power received at users connecting to macro BS ℓ (where $\ell \in \Phi_M$) is given by

$$P_{r,1} = G_M P_1 L\left(d_{\ell,1}\right)/N, \qquad (3.2)$$

where P_1 is the transmit power of a macro BS, $L\left(d_{\ell,1}\right) = \eta d_{\ell,1}^{-\alpha_1}$ is large-scale path loss, $d_{\ell,1}$ is the distance between the user and a macro BS, η is the frequency dependent factor, and B_i is the identical bias factor which is useful for offloading data traffic in HetNets.

3.1.1.3 Channel Model

In small cells, without loss of generality, we consider that each small cell BS is associated with one user in the previous round of user association process. With applying NOMA protocol, we aim to squeeze a typical user into a same small cell to improve the spectral efficiency. For simplicity, we assume that the distances between the associated users and the connected small cell BSs are the same, which can be arbitrary values and are denoted as r_k; future work will relax this assumption. The distance between atypical user and the connected small cell BS is random. Due to the fact that the path loss is more stable and dominant compared to the instantaneous small-scale fading, we assume that the SIC operation always happened at the near user. We denote that d_{o,k_m} and d_{o,k_n} are the distances from the k-th tier small cell BS to user m and user n, respectively. Since it is not predetermined that atypical user is a near user n or a far user m, we have the following cases.

Near User Case When atypical user is a near user n ($x \leq r_k$), then we have $d_{o,k_m} = r_k$. User n will first decode the information of the connected user m^* to the same BS with the following SINR

$$\gamma_{k_{n \to m^*}} = \frac{a_{m,k} P_k g_{o,k} L\left(d_{o,k_n}\right)}{a_{n,k} P_k g_{o,k} L\left(d_{o,k}\right) + I_{M,k} + I_{S,k} + \sigma^2}, \qquad (3.3)$$

where $a_{m,k}$ and $a_{n,k}$ are the power sharing coefficients for two users in the k-th layer, σ^2 is the additive white Gaussian noise (AWGN) power, $L\left(d_{o,k_n}\right) = \eta d_{o,k_n}^{-\alpha_i}$ is the large-scale path loss, $I_{M,k} = \sum_{\ell \in \Phi_1} \frac{P_1}{N} g_{\ell,1} L\left(d_{\ell,1}\right)$ is the interference from macro cells, $I_{S,k} = \sum_{i=2}^{K} \sum_{j \in \Phi_i \setminus B_{o,k}} P_i g_{j,i} L\left(d_{j,i}\right)$ is the interference from small cells,

$g_{o,k}$ and d_{o,k_n} refer the small-scale fading coefficients and distance between atypical user and the connected BS in the k-th tier, $g_{\ell,1}$ and $d_{\ell,1}$ refer the small-scale fading coefficients and distance between a typical user and connected BS ℓ in the macro cell, respectively, $g_{j,i}$ and $d_{j,i}$ refer the small-scale fading coefficients and distance between a typical user and its connected BS j except the serving BS $B_{o,k}$ in the i-th tier small cell, respectively. Here, $g_{o,k}$ and $g_{j,i}$ follow exponential distributions with unit mean. $g_{\ell,1}$ follows Gamma distribution with parameters $(N, 1)$.

If the information of user m^* can be decoded successfully, user n then decodes its own message. As such, the SINR at atypical user n, which connects with the k-th tier small cell, can be expressed as

$$\gamma_{k_n} = \frac{a_{n,k} P_k g_{o,k} L\left(d_{o,k_n}\right)}{I_{M,k} + I_{S,k} + \sigma^2}. \tag{3.4}$$

For the connected far user m^* to the same BS, the signal can be decoded by treating the message of user n as interference. Therefore, the SINR that for the connected user m^* to the same BS in the k-th tier small cell can be expressed as

$$\gamma_{k_{m^*}} = \frac{a_{m,k} P_k g_{o,k} L\left(r_k\right)}{I_{k,n} + I_{M,k} + I_{S,k} + \sigma^2}, \tag{3.5}$$

where $I_{k,n} = a_{n,k} P_k g_{o,k} L\left(r_k\right)$, and $L\left(r_k\right) = \eta r_k^{-\alpha_k}$.

Far User Case When atypical user is the far user m $(x > r_k)$, we have $d_{o,k_n} = r_k$. As such, for the connected near user n^*, it will first decode the information of user m with the following SINR

$$\gamma_{k_{n^* \to m}} = \frac{a_{m,k} P_k g_{o,k} L\left(r_k\right)}{a_{n,k} P_k g_{o,k} L\left(r_k\right) + I_{M,k} + I_{S,k} + \sigma^2}. \tag{3.6}$$

Once user m is decoded successfully, the interference from atypical user m can be canceled, by applying the SIC technology. Therefore, the SINR at the connected user n^* to the same BS in the k-th tier small cell is given by

$$\gamma_{k_{n^*}} = \frac{a_{n,k} P_k g_{o,k} L\left(r_k\right)}{I_{M,k} + I_{S,k} + \sigma^2}. \tag{3.7}$$

For user m that connects to the k-th tier small cell, the SINR can be expressed as

$$\gamma_{k_m} = \frac{a_{m,k} P_k g_{o,k} L\left(d_{o,k_m}\right)}{I_{k,n^*} + I_{M,k} + I_{S,k} + \sigma^2}, \tag{3.8}$$

where $I_{k,n^*} = a_{n,k} P_k g_{o,k} L\left(d_{o,k_m}\right)$, $L\left(d_{o,k_m}\right) = \eta d_{o,k_m}^{-\alpha_k}$, d_{o,k_n} is the distance between atypical user m and the connected BS in the k-th tier.

Without loss of generality, we assume that a typical user is located at the origin of an infinite two-dimensional plane. Based on (3.1) and (3.2), the SINR at atypical user that connects with a macro BS at a random distance $d_{o,1}$ can be expressed as

$$\gamma_{r,1} = \frac{\frac{P_1}{N}h_{o,1}L\left(d_{o,1}\right)}{I_{M,1} + I_{S,1} + \sigma^2},$$

(3.9)

where $I_{M,1} = \sum_{\ell \in \Phi_1 \setminus B_{o,1}} \frac{P_1}{N}h_{\ell,1}L\left(d_{\ell,1}\right)$ is the interference from macro cells, $I_{S,1} = \sum_{i=2}^{K}\sum_{j\in\Phi_i} P_i h_{j,i}L\left(d_{j,i}\right)$ is the interference from small cells, $h_{o,1}$ is the small-scale fading coefficient between atypical user and the connected macro BS, $h_{\ell,1}$ and $d_{\ell,1}$ refer the small-scale fading coefficients and distance between a typical user and the connected macro BS ℓ except the serving BS $B_{o,1}$ in the macro cell, respectively, $h_{j,i}$ and $d_{j,i}$ refer the small-scale fading coefficients and distance between atypical user and connected BS j in the i-th tier small cell, respectively. Here, $h_{o,1}$ follows Gamma distribution with parameters $(M - N + 1, 1)$, $h_{\ell,1}$ follows Gamma distribution with parameters $(N, 1)$, and $h_{j,i}$ follows exponential distribution with unit mean.

3.1.2 Coverage Probability of Non-orthogonal Multiple-Access-Based Small Cells

In this subsection, we focus our attention on analyzing the coverage probability of a typical user associated with the NOMA enhanced small cells, which is different from the conventional OMA-based small cells due to the channel ordering of two users. The analysis of coverage probability of a typical user associated with the massive MIMO-aided macro cells is the same as the conventional massive MIMO-aided OMA small cells.

3.1.2.1 User Association Probability and Distance Distributions

The user association of the proposed framework is based on maximizing the biased average received power at users. As such, based on (3.1) and (3.2), the user association of macro cells and small cells is given in the following. For simplicity, we denote $\tilde{B}_{ik} = \frac{B_i}{B_k}$, $\tilde{\alpha}_{ik} = \frac{\alpha_i}{\alpha_k}$, $\tilde{\alpha}_{1k} = \frac{\alpha_1}{\alpha_k}$, $\tilde{\alpha}_{i1} = \frac{\alpha_i}{\alpha_1}$, $\tilde{P}_{1k} = \frac{P_1}{P_k}$, $\tilde{P}_{i1} = \frac{P_i}{P_1}$, and $\tilde{P}_{ik} = \frac{P_i}{P_k}$ in the following parts of this treatise.

Lemma 3.1 *The user association probability that a typical user connects to NOMA enhanced small cell BSs in the k-th tier and to macro BSs can be calculated as*

$$A_k = 2\pi\lambda_k \int_0^\infty r \exp\left[-\pi\sum_{i=2}^{K}\lambda_i\left(\tilde{P}_{ik}\tilde{B}_{ik}\right)^{\delta_i} r^{\frac{2}{\tilde{\alpha}_{ik}}} - \pi\lambda_1\left(\frac{\tilde{P}_{1k}G_M}{Na_{n,k}B_k}\right)^{\delta_1} r^{\frac{2}{\tilde{\alpha}_{1k}}}\right] dr.,$$

(3.10)

and

$$A_1 = 2\pi\lambda_1 \int_0^\infty r \exp\left[-\pi \sum_{i=2}^K \lambda_i \left(\frac{a_{n,i}\tilde{P}_{i1}B_i N}{G_M}\right)^{\delta_i} r^{\frac{2}{\alpha_{i1}}} - \pi\lambda_1 r^2\right] dr., \quad (3.11)$$

respectively, where $\delta_1 = \frac{2}{\alpha_1}$ and $\delta_i = \frac{2}{\alpha_i}$.

Proof *Using the similar method as Lemma 1 of Jo et al. (2012), (3.10) and (3.11) can be easily obtained.*

Corollary 3.1 *For the special case that each tier has the same path loss exponent, i.e., $\alpha_1 = \alpha_k = \alpha$, the user association probability of the NOMA enhanced small cells in the k-th tier and macro cells can be expressed in closed form as*

$$\tilde{A}_k = \frac{\lambda_k}{\sum_{i=2}^K \lambda_i \left(\tilde{P}_{ik}\tilde{B}_{ik}\right)^\delta + \lambda_1\left(\frac{\tilde{P}_{1k}G_M}{Na_{n,k}B_k}\right)^\delta}, \quad (3.12)$$

and

$$\tilde{A}_1 = \frac{\lambda_1}{\sum_{i=2}^K \lambda_i \left(\frac{a_{n,i}\tilde{P}_{i1}B_i N}{G_M}\right)^\delta + \lambda_1}, \quad (3.13)$$

respectively, where $\delta = \frac{2}{\alpha}$.

Remark 3.1 The derived results in (3.12) and (3.13) demonstrate that by increasing the number of antennas at the macro cell BSs, the user association probability of the macro cells increases and the user association probability of the small cells decreases. This is due to the large array gains brought by the macro cells to the served users. It is also worth noting that increasing the power sharing coefficient, a_n, results in higher association probability of small cells. As $a_n \to 1$, the user association becomes the same as in the conventional OMA-based approach.

Then we consider the probability density function (PDF) of the distance between a typical user and its connected small cell BS in the k-th tier. Based on (3.10), we obtain

$$f_{d_{o,k}}(x) = \frac{2\pi\lambda_k x}{A_k} \exp\left[-\pi \sum_{i=2}^K \lambda_i \left(\tilde{P}_{ik}\tilde{B}_{ik}\right)^{\delta_i} x^{\frac{2}{\alpha_{ik}}} - \pi\lambda_1\left(\frac{\tilde{P}_{1k}G_M}{Na_{n,k}B_k}\right)^{\delta_1} x^{\frac{2}{\alpha_{1k}}}\right].$$

$$(3.14)$$

We then calculate the PDF of the distance between a typical user and its connected macro BS. Based on (3.11), we obtain

$$f_{d_{o,1}}(x) = \frac{2\pi \lambda_1 x}{A_1} \exp\left[-\pi \sum_{i=2}^{K} \lambda_i \left(\frac{a_{n,i} \tilde{P}_{i1} B_i N}{G_M}\right)^{\delta_i} x^{\frac{2}{\tilde{\alpha}_{i1}}} - \pi \lambda_1 x^2\right]. \tag{3.15}$$

3.1.2.2 Laplace Transform of Interferences

The next step is to derive the Laplace transform of a typical user. We denote that $I_k = I_{S,k} + I_{M,k}$ is the total interference to the typical user in the k-th tier. The Laplace transform of I_k is $\mathscr{L}_{I_k}(s) = \mathscr{L}_{I_{S,k}}(s)\mathscr{L}_{I_{M,k}}(s)$. We first calculate the Laplace transform of interference from the small cell BS to a typical user $\mathscr{L}_{I_{S,k}}(s)$ in the following lemma.

Lemma 3.2 *The Laplace transform of interferences from the small cell BSs to a typical user can be expressed as*

$$\mathscr{L}_{I_{S,k}}(s) = \exp\left\{-s \sum_{i=2}^{K} \frac{\lambda_i 2\pi P_i \eta (\omega_{i,k}(x_0))^{2-\alpha_i}}{\alpha_i (1-\delta_i)}\right.$$

$$\left. \times {}_2F_1\left(1, 1-\delta_i; 2-\delta_i; -s P_i \eta (\omega_{i,k}(x_0))^{-\alpha_i}\right)\right\}, \tag{3.16}$$

where ${}_2F_1(\cdot, \cdot; \cdot; \cdot)$ is the Gauss hypergeometric function (Gradshteyn and Ryzhik 2000, Eq. (9.142)), and $\omega_{i,k}(x_0) = \left(\tilde{B}_{ik}\tilde{P}_{ik}\right)^{\frac{\delta_i}{2}} x_0^{\frac{1}{\tilde{\alpha}_{ik}}}$ is the nearest distance allowed between the typical user and its connected small cell BS in the k-th tier.

Then we calculate the Laplace transform of interference from the macro cell to a typical user $\mathscr{L}_{I_{M,k}}(s)$ in the following lemma.

Lemma 3.3 *The Laplace transform of interference from the macro cell BSs to a typical user can be expressed as*

$$\mathscr{L}_{I_{M,k}}(s) = \exp\left[-\lambda_1 \pi \delta_1 \sum_{p=1}^{N} \binom{N}{p}\left(s\frac{P_1}{N}\eta\right)^p \left(-s\frac{P_1}{N}\eta\right)^{\delta_1 - p}\right.$$

$$\left. \times B\left(-s\frac{P_1}{N}\eta [\omega_{1,k}(x_0)]^{-\alpha_1}; p-\delta_1, 1-N\right)\right], \tag{3.17}$$

where $B(\cdot; \cdot, \cdot)$ is the incomplete Beta function (Gradshteyn and Ryzhik 2000, Eq. (8.319)), and $\omega_{1,k}(x_0) = \left(\frac{\tilde{P}_{1k} G_M}{a_{n,k} B_k N}\right)^{\frac{\delta_1}{2}} x^{\frac{1}{\alpha_{1k}}}$ is the nearest distance allowed between a typical user and its connected BS in the macro cell.

3.1.2.3 Coverage Probability

The coverage probability is defined as that a typical user can successfully transmit signals with targeted data rate R_t. According to the distances, two cases are considered in the following.

Near User Case For the near user case, $x_0 < r_k$, the success decoding will happen when the following two conditions hold:

1. The typical user can decode the message of the connected user served by the same BS.
2. After the SIC process, the typical user can decode its own message.

As such, the coverage probability of the typical user on the condition of the distance x_0 in the k-th tier is:

$$P_{cov,k}(\tau_c, \tau_t, x_0)\big|_{x_0 \leq r_k} = \Pr\left\{ \gamma_{k_n \to m_*} > \tau_c, \gamma_{k_n} > \tau_t \right\}, \tag{3.18}$$

where $\tau_t = 2^{R_t} - 1$ and $\tau_c = 2^{R_c} - 1$. Here R_c is the targeted data rate of the connected user served by the same BS.

Based on (3.18), for the near user case, we can obtain the expressions for the conditional coverage probability of a typical user in the following lemma.

Lemma 3.4 *If $a_{m,k} - \tau_c a_{n,k} \geq 0$ holds, the conditional coverage probability of a typical user for the near user case is expressed in closed form as*

$$
\begin{aligned}
P_{cov,k}(\tau_c, \tau_t, x_0)\big|_{x_0 \leq r_k} = \exp\bigg\{ & -\frac{\varepsilon^*(\tau_c, \tau_t) x_0^{\alpha_k} \sigma^2}{P_k \eta} \\
& - \lambda_1 \delta_1 \pi \left(\tilde{P}_{1k} \varepsilon^*(\tau_c, \tau_t) / N \right)^{\delta_1} x_0^{\frac{2}{\alpha_{1k}}} Q_{1,t}^n(\tau_c, \tau_t) \\
& - \sum_{i=2}^{K} \frac{\lambda_i \delta_i \pi \left(\tilde{B}_{ik} \right)^{\frac{2}{\alpha_i} - 1} \left(\tilde{P}_{ik} \right)^{\frac{2}{\alpha_i}} x_0^{\frac{2}{\alpha_{ik}}}}{1 - \delta_i} Q_{i,t}^n(\tau_c, \tau_t) \bigg\}.
\end{aligned}
\tag{3.19}
$$

Otherwise, $P_{cov,k}(\tau_c, \tau_t, x_0)\big|_{x_0 \leq r_k} = 0.$ *Here,* $\varepsilon_t^n = \frac{\tau_t}{a_{n,k}}$, $\varepsilon_c^f = \frac{\tau_c}{a_{m,k} - \tau_c a_{n,k}}$, $\varepsilon^*(\tau_c, \tau_t) = \max\left\{ \varepsilon_c^f, \varepsilon_t^n \right\}$, $Q_{i,t}^n(\tau_c, \tau_t) = \varepsilon^*(\tau_c, \tau_t) \, {}_2F_1\left(1, 1 - \delta_i; 2 - \delta_i; -\frac{\varepsilon^*(\tau_c, \tau_t)}{\tilde{B}_{ik}} \right)$, *and* $Q_{1,t}^n(\tau_c, \tau_t) = \sum_{p=1}^{N} \binom{N}{p} (-1)^{\delta_1 - p} \times B\left(-\frac{\varepsilon^*(\tau_c, \tau_t) a_{n,k} B_k}{G_M} \right.$; $p - \delta_1, 1 - N \bigg).$

Far User Case For the far user case, $x_0 > r_k$, the success decoding will happen if the typical user can decode its own message by treating the connected user served by the same BB as noise. The conditional coverage probability of a typical user for far user case is calculated in the following lemma.

Lemma 3.5 *If $a_{m,k} - \tau_t a_{n,k} \geq 0$ holds, the coverage probability of a typical user for the far user case is expressed in closed form as*

$$
P_{cov,k}\left(\tau_t, x_0\right)\big|_{x_0 > r_k} = \exp\left\{ -\frac{\varepsilon_t^f x_0^{\alpha_k} \sigma^2}{P_k \eta} - \lambda_1 \delta_1 \pi \left(\tilde{P}_{1k}\varepsilon_t^f / N\right)^{\delta_1} x_0^{\frac{2}{\tilde{\alpha}_{1k}}} Q_{1,t}^f\left(\tau_t\right) \right.
$$

$$
\left. - \sum_{i=2}^{K} \frac{\lambda_i \delta_i \pi \left(\tilde{B}_{ik}\right)^{\frac{2}{\alpha_i}-1} \left(\tilde{P}_{ik}\right)^{\frac{2}{\alpha_i}} x_0^{\frac{2}{\tilde{\alpha}_{ik}}}}{1 - \delta_i} Q_{i,t}^f\left(\tau_t\right) \right\}. \qquad (3.20)
$$

Otherwise, $P_{cov,k}\left(\tau_t, x_0\right)\big|_{x_0 > r_k} = 0.$ Here $\varepsilon_t^f = \frac{\tau_t}{a_{m,k} - \tau_t a_{n,k}}$, and

$$
Q_{1,t}^f\left(\tau_t\right) = \sum_{p=1}^{N} \binom{N}{p}(-1)^{\delta_1 - p} B\left(-\frac{\varepsilon_t^f a_{n,k} B_k}{G_M}; p - \delta_1, 1 - N\right)
$$

$$
Q_{i,t}^f\left(\tau_t\right) = \varepsilon_t^f {}_2F_1\left(1, 1 - \delta_i; 2 - \delta_i; -\frac{\varepsilon_t^f}{B_{ik}}\right).
$$

Based on Lemmas 3.4 and 3.5, we can calculate the coverage probability of a typical user in the following theorem.

Theorem 3.1 *The coverage probability of a typical user associated with the k-th tier small cells is expressed as*

$$
P_{cov,k}\left(\tau_c, \tau_t\right) = \int_0^{r_k} P_{cov,k}\left(\tau_c, \tau_t, x_0\right)\big|_{x_0 \leq r_k} f_{do,k}\left(x_0\right) dx_0
$$

$$
+ \int_{r_k}^{\infty} P_{cov,k}\left(\tau_t, x_0\right)\big|_{x_0 > r_k} f_{do,k}\left(x_0\right) dx_0, \qquad (3.21)
$$

where $P_{cov,k}\left(\tau_c, \tau_t, x_0\right)\big|_{x_0 \leq r_k}$ is given in (3.19), $P_{cov,k}\left(\tau_t, x_0\right)\big|_{x_0 > r_k}$ is given in (3.20), and $f_{do,k}\left(x_0\right)$ is given in (3.14).

Although (3.21) has provided the exact analytical expression for the coverage probability of typical user, it is hard to directly obtain insights from this expression. Driven by this, we provide one special case with considering that each tier is with the same path loss exponents. As such, we have $\tilde{\alpha}_{1k} = \tilde{\alpha}_{ik} = 1$. In addition, we consider the interference limited case, where the thermal noise can be neglected. Then based on (3.21), we can obtain the closed-form coverage probability of a typical user in the following corollary.

Corollary 3.2 *With $\alpha_1 = \alpha_k = \alpha$ and $\sigma^2 = 0$, the coverage probability of a typical user can be expressed in closed form as follows:*

$$\tilde{P}_{cov,k}\left(\tau_c, \tau_t\right) = \frac{b_k \left(1 - e^{-\pi \left(b_k + c_1^n(\tau_c, \tau_t) + c_2^n(\tau_c, \tau_t)\right) r_k^2}\right)}{b_k + c_1^n\left(\tau_c, \tau_t\right) + c_2^n\left(\tau_c, \tau_t\right)} + \frac{b_k e^{-\pi\left(b_k + c_1^f(\tau_t) + c_2^f(\tau_t)\right) r_k^2}}{b_k + c_1^f\left(\tau_t\right) + c_2^f\left(\tau_t\right)},$$
$$\text{(3.22)}$$

where $b_k = \sum_{i=2}^{K} \lambda_i \left(\tilde{P}_{ik} \tilde{B}_{ik}\right)^\delta + \lambda_1 \left(\frac{\tilde{P}_{1k} G_M}{N a_{n,k} B_k}\right)^\delta$, $c_1^n\left(\tau_c, \tau_t\right) = \lambda_1 \delta_1 \left(\frac{\tilde{P}_{1k} \varepsilon^*(\tau_c, \tau_t)}{N}\right)^\delta \tilde{Q}_{1,t}^n$

(τ_c, τ_t), $c_2^n\left(\tau_c, \tau_t\right) = \sum_{i=2}^{K} \frac{\lambda_i \delta_i \left(\tilde{B}_{ik}\right)^{\frac{2}{\alpha}-1} \left(\tilde{P}_{ik}\right)^{\frac{2}{\alpha}}}{1-\delta_i} \tilde{Q}_{i,t}^n\left(\tau_c, \tau_t\right)$, $c_1^f\left(\tau_t\right) = \lambda_1 \delta_1 \left(\frac{\tilde{P}_{1k}\varepsilon_t^f}{N}\right)^{\delta_1}$

$\tilde{Q}_{1,t}^f\left(\tau_t\right)$, and $c_2^f\left(\tau_t\right) = \sum_{i=2}^{K} \frac{\lambda_i \delta_i \left(\tilde{B}_{ik}\right)^{\frac{2}{\alpha}-1} \left(\tilde{P}_{ik}\right)^{\frac{2}{\alpha}}}{1-\delta} \tilde{Q}_{i,t}^f\left(\tau_t\right)$. *Here,* $\tilde{Q}_{1,t}^n\left(\tau_c, \tau_t\right)$, $\tilde{Q}_{i,t}^n$

(τ_c, τ_t), $\tilde{Q}_{1,t}^f\left(\tau_t\right)$, *and* $\tilde{Q}_{i,t}^f\left(\tau_t\right)$ *are based on interchanging the same path loss expo-nents, i.e.,* $\alpha_1 = \alpha_k = \alpha$, *for each tier from* $Q_{1,t}^n\left(\tau_c, \tau_t\right)$, $Q_{i,t}^n\left(\tau_c, \tau_t\right)$, $Q_{1,t}^f\left(\tau_t\right)$, *and* $Q_{i,t}^f\left(\tau_t\right)$.

Remark 3.2 The derived results in (3.22) demonstrate that the coverage probability of a typical user is determined by the target rate of itself as well as the target rate of the connected user. Additionally, inappropriate power allocation such as, $a_{m,k} - \tau_t a_{n,k} < 0$, will lead to the coverage probability always being zero.

3.1.3 Spectrum Efficiency

To evaluate the spectrum efficiency of the proposed NOMA enhanced hybrid HetNets framework, we calculate the spectrum efficiency of each tier in this section.

3.1.3.1 Ergodic Rate of NOMA Enhanced Small Cells

Different from calculating the coverage probability of the case with fixed targeted rate, the achievable ergodic rate for NOMA enhanced small cells is opportunistically determined by the channel conditions of users. It is also easy to verify that if the far user can decode the message of itself, the near user can definitely decode the message of far user since it has a better channel condition (Ding et al. 2014). Recall that the distance order between the connected BS and the two users are not predetermined, as such, we calculate the achievable ergodic rate of small cells for both the near user case and far user case in the following lemmas.

Lemma 3.6 *The achievable ergodic rate of the k-th tier small cell for the near user case can be expressed as follows:*

$$
\tau_k^n = \frac{2\pi\lambda_k}{A_k \ln 2} \left[\int_0^{\frac{a_{m,k}}{a_{n,k}}} \frac{\bar{F}_{\gamma_{km*}}(z)}{1+z} dz + \int_0^\infty \frac{\bar{F}_{\gamma_{kn}}(z)}{1+z} dz \right], \tag{3.23}
$$

where $\bar{F}_{\gamma_{km*}}(z)$ and $\bar{F}_{\gamma_{kn}}(z)$ are given by

$$
\bar{F}_{\gamma_{km*}}(z) = \int_0^{r_k} x \exp\left[-\frac{\sigma^2 z r_k^{\alpha_k}}{\left(a_{m,k} - a_{n,k}z\right) P_k \eta} \right.
$$
$$
\left. -\Theta\left(\frac{z r_k^{\alpha_k}}{\left(a_{m,k} - a_{n,k}z\right) P_k \eta} \right) + \Lambda(x) \right] dx, \tag{3.24}
$$

and

$$
\bar{F}_{\gamma_{kn}}(z) = \int_0^{r_k} x \exp\left[\Lambda(x) - \frac{\sigma^2 z x^{\alpha_k}}{a_{n,k} P_k \eta} - \Theta\left(\frac{z x^{\alpha_k}}{a_{n,k} P_k \eta} \right) \right] dx. \tag{3.25}
$$

Here $\Lambda(x) = -\pi \sum_{i=2}^K \lambda_i \left(\tilde{P}_{ik} \tilde{B}_{ik} \right)^{\delta_i} x^{\frac{2}{\alpha_{ik}}} - \pi \lambda_1 \left(\frac{\tilde{P}_{1k} G_M}{N a_{n,k} B_k} \right)^{\delta_1} x^{\frac{2}{\alpha_{1k}}}$ and $\Theta(s)$ is given by

$$
\Theta(s) = \lambda_1 \pi \delta_1 \sum_{p=1}^N \binom{N}{p} \left(s \frac{P_1}{N} \eta \right)^p \left(-s \frac{P_1}{N} \eta \right)^{\delta_1 - p}
$$
$$
\times B\left(-s\frac{P_1}{N} \eta \left[\omega_{1,k}(x)\right]^{-\alpha_1}; p - \delta_1, 1 - N \right)
$$
$$
+ s \sum_{i=2}^K \frac{\lambda_i 2\pi P_i \eta \left(\omega_{i,k}(x) \right)^{2-\alpha_i}}{\alpha_i (1 - \delta_i)}
$$
$$
\times {}_2F_1\left(1, 1 - \delta_i; 2 - \delta_i; -s P_i \eta \left(\omega_{i,k}(x) \right)^{-\alpha_i} \right). \tag{3.26}
$$

Lemma 3.7 *The achievable ergodic rate of the k-th tier small cell for the far user case can be expressed as follows:*

$$
\tau_k^f = \frac{2\pi\lambda_k}{A_k \ln 2} \left[\int_0^\infty \frac{\bar{F}_{\gamma_{kn*}}(z)}{1+z} dz + \int_0^{\frac{a_{m,k}}{a_{n,k}}} \frac{\bar{F}_{\gamma_{km}}(z)}{1+z} dz \right], \tag{3.27}
$$

where $\bar{F}_{\gamma_{km}}(z)$ and $\bar{F}_{\gamma_{kn*}}(z)$ are given by

$$\bar{F}_{\gamma_{km}}(z) = \int_{r_k}^{\infty} x \exp\left[-\frac{\sigma^2 z x^{\alpha_k}}{P_k \eta \left(a_{m,k} - a_{n,k} z\right)}\right.$$

$$\left.-\Theta\left(\frac{z x^{\alpha_k}}{P_k \eta \left(a_{m,k} - a_{n,k} z\right)}\right) + \Lambda(x)\right] dx, \tag{3.28}$$

and

$$\bar{F}_{\gamma_{kn*}}(z) = \int_{r_k}^{\infty} x \exp\left[\Lambda(x) - \frac{\sigma^2 z r_k^{\alpha_k}}{P_k \eta a_{n,k}} - \Theta\left(\frac{z r_k^{\alpha_k}}{P_k \eta a_{n,k}}\right)\right] dx. \tag{3.29}$$

Theorem 3.2 *Conditioned on the HPPPs, the achievable ergodic rate of the small cells can be expressed as follows:*

$$\tau_k = \tau_k^n + \tau_k^f, \tag{3.30}$$

where τ_k^n and τ_k^f are obtained from (3.23) and (3.27).

Note that the derived results in (3.30) is a double integral form, since even for some special cases, it is challenging to obtain closed form solutions. However, the derived expression is still much more efficient and also more accurate compared to using the approach of Monte Carlo simulations, which highly depends on the repeated iterations of random sampling.

3.1.3.2 Ergodic Rate of Macro Cells

In massive MIMO-aided macro cells, the achievable ergodic rate can be significantly improved due to multiple-antenna array gains, but with more power consumption and high complexity. However, the exact analytical results require high order derivatives of Laplace transform. When the number of antennas goes large, it becomes mathematical intractable to calculate the derivatives due to the unacceptable complexity. In order to evaluate the spectrum efficiency of the whole system, we provide a tractable lower bound of throughput for macro cells in the following theorem.

Theorem 3.3 *The lower bound of achievable ergodic rate of the macro cells can be expressed as follows:*

$$\tau_{1,L} = \log_2\left(1 + \frac{P_1 G_M \eta}{N \int_0^{\infty} \left(Q_1(x) + \sigma^2\right) x^{\alpha_1} f_{d_{o,1}}(x)\, dx}\right), \tag{3.31}$$

where $f_{d_{o,1}}(x)$ is given in (3.15), $Q_1(x) = \frac{2P_1\eta\pi\lambda_1}{\alpha_1-2}x^{2-\alpha_1} + \sum_{i=2}^{K} 2\pi\lambda_i \left(\frac{P_i\eta}{\alpha_i-2}\right)$ $[\omega_{i,1}(x)]^{2-\alpha_i}$, and $\omega_{i,1}(x) = \left(\frac{a_{n,i}\tilde{P}_{i1}B_iN}{G_M}\right)^{\frac{\delta_i}{2}} x^{\frac{1}{\alpha_{i1}}}$ is denoted as the nearest distance allowed between the i-th tier small cell BS and the typical user that is associated with the macro cell.

Corollary 3.3 *If $\alpha_1 = \alpha_k = \alpha$ holds, the lower bound of achievable ergordic rate of the macro cell is given by in closed form as*

$$\tilde{\tau}_{1,L} = \log_2\left(1 + \frac{P_1 G_M \eta/N}{\psi(\pi b_1)^{-1} + \sigma^2 \Gamma\left(\frac{\alpha}{2}+1\right)(\pi b_1)^{-\frac{\alpha}{2}}}\right), \tag{3.32}$$

where $\psi = \frac{2P_1\eta\pi\lambda_1}{\alpha-2} + \sum_{i=2}^{K}\left(\frac{2\pi\lambda_i P_i\eta}{\alpha-2}\right)\left(\frac{a_{n,i}\tilde{P}_{i1}B_iN}{G_M}\right)^{\delta-1}$ *and* $b_1 = \sum_{i=2}^{K}\lambda_i\left(\frac{a_{n,i}\tilde{P}_{i1}B_iN}{G_M}\right)^{\delta} +$ λ_1.

Remark 3.3 The derived results in (3.32) demonstrate that achievable ergordic rate of the macro cell can be enhanced by increasing the number of antennas at the macro cell BSs. This is because the users in macro cells can experience larger array gains.

3.1.3.3 Spectrum Efficiency of the Proposed Hybrid Hetnets

Based on the analysis of last two subsections, a tractable lower bound of spectrum efficiency can be given in the following proposition.

Proposition 3.1 *The spectrum efficiency of the proposed hybrid Hetnets is*

$$\tau_{SE,L} = A_1 N \tau_{1,L} + \sum_{k=2}^{K} A_k \tau_k, \tag{3.33}$$

where $N\tau_1$ and τ_k are the low bound spectrum efficiency of macro cells and exact spectrum efficiency of the k-th tier small cells, respectively. Here, A_k and A_1 are obtained from (3.10) and (3.11), and τ_k and $\tau_{1,L}$ are obtained from (3.30) and (3.31), respectively.

3.1.4 Energy Efficiency

In this section, we proceed to investigate the performance of the proposed hybrid HetNets framework from the perspective of energy efficiency, due to the fact that energy efficiency is an important performance metric in the 5G systems.

3.1.4.1 Power Consumption Model

To calculate the energy efficiency of the considered networks, we first need to model the power consumption parameter of both small cell BSs and macro cell BSs. The power consumption of small cell BSs is given by

$$P_{i,total} = P_{i,static} + \frac{P_i}{\varepsilon_i}, \tag{3.34}$$

where $P_{i,static}$ is the static hardware power consumption of small cell BSs in the i-th tier, and ε_i is the efficiency factor for the power amplifier of small cell BSs in the i-th tier.

The power consumption of macro cell BSs is given by

$$P_{1,total} = P_{1,static} + \sum_{a=1}^{3} \left(N^a \Delta_{a,0} + N^{a-1} M \Delta_{a,1} \right) + \frac{P_1}{\varepsilon_1}, \tag{3.35}$$

where $P_{1,static}$ is the static hardware power consumption of macro cell BSs, ε_1 is the efficiency factor for the power amplifier of macro cell BSs, and $\Delta_{a,0}$ and $\Delta_{a,1}$ are the practical parameters which depended on the chains of transceivers, precoding, coding/decoding, etc.

3.1.4.2 Energy Efficiency of NOMA Enhanced Small Cells and Macro Cells

The energy efficiency is defined as

$$\Theta_{EE} = \frac{\text{Total data rate}}{\text{Total energy consumption}}. \tag{3.36}$$

Therefore, based on (3.36) and the power consumption model for small cells that we have provided in (3.34), the energy efficiency of the k-th tier of NOMA enhanced small cells is expressed as

$$\Theta_{EE}^k = \frac{\tau_k}{P_{k,total}}, \tag{3.37}$$

where τ_k is obtained from (3.30).

Based on (3.35) and (3.36), the energy efficiency of macro cell is expressed as

$$\Theta_{EE}^1 = \frac{N\tau_{1,L}}{P_{1,total}}, \tag{3.38}$$

where $\tau_{1,L}$ is obtained from (3.31).

3.1.4.3 Energy Efficiency of the Proposed Hybrid Hetnets

According to the derived results of energy efficiency of NOMA enhanced small cells and macro cells, we can express the energy efficiency in the following proposition.

Proposition 3.2 *The energy efficiency of the proposed hybrid Hetnets is as follows:*

$$\Theta_{\text{EE}}^{\text{Hetnets}} = A_1 \Theta_{\text{EE}}^1 + \sum_{k=2}^{K} A_k \Theta_{\text{EE}}^k, \tag{3.39}$$

where A_k and A_1 are obtained from (3.10) and (3.11), and Θ_{EE}^k and Θ_{EE}^1 are obtained from (3.37) and (3.38).

3.1.5 Numerical Results

In this section, numerical results are presented to facilitate the performance evaluations of NOMA enhanced hybrid K-tier HetNets. The noise power is $\sigma^2 = -170 + 10 \times \log_{10}(BW) + N_f$. The power sharing coefficients of NOMA for each tier are same as $a_{m,k} = a_m$ and $a_{n,k} = a_n$ for simplicity. BPCU is short for bit per channel use. Monte Carlo simulations marked as "o" are provided to verify the accuracy of our analysis. Table 3.1 summarizes the simulation parameters used in this section.

3.1.5.1 User Association Probability and Coverage Probability

Figure 3.2 shows the effect of number of antennas equipped at each macro BS, M, and bias factor on user association probability, where the tiers of HetNets are set to

Table 3.1 Table of parameters

Monte Carlo simulations repeated	10^5 times
The radius of the plane	10^4 m
Carrier frequency	1 GHz
The BS density of macro cells	$\lambda_1 = \left(500^2 \times \pi\right)^{-1}$
Pass loss exponent	$\alpha_1 = 3.5, \alpha_k = 4$
The noise figure	$N_f = 10$ dB
The noise power	$\sigma^2 = -90$ dBm
Static hardware power consumption	$P_{1,total} = 4$ W, $P_{i,total} = 2$ W
Power amplifier efficiency factor	$\varepsilon_1 = \varepsilon_i = 0.4$
Precoding power consumption	$\Delta_{1,0} = 4.8, \Delta_{2,0} = 0$
	$\Delta_{3,0} = 2.08 \times 10^{-8}$
	$\Delta_{1,1} = 1, \Delta_{2,1} = 9.5 \times 10^{-8}$
	$\Delta_{3,1} = 6.25 \times 10^{-8}$

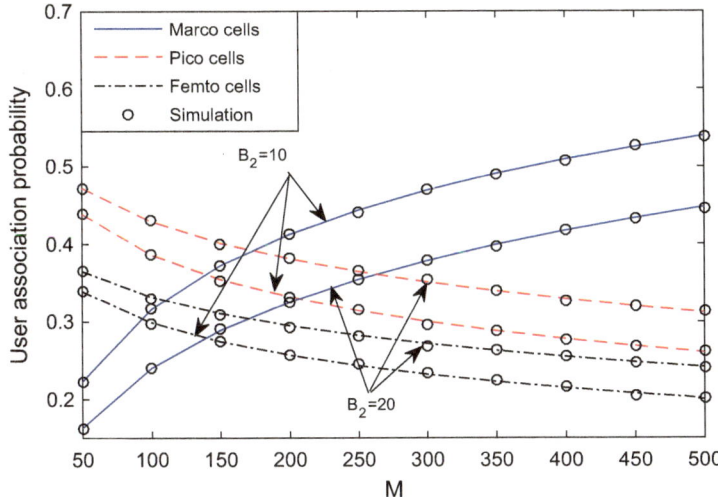

Fig. 3.2 User association probability versus antenna number with different bias factor, with $K = 3$, $N = 15$, $P_1 = 40\,\text{dBm}$, $P_2 = 30\,\text{dBm}$ and $P_3 = 20\,\text{dBm}$, $r_k = 50\,\text{m}$, $a_m = 0.6$, $a_n = 0.4$, $\lambda_2 = \lambda_3 = 20 \times \lambda_1$, and $B_3 = 20 \times B_2$

be $K = 3$, including macro cells and two tiers of small cells. The analytical curves representing small cells and macro cells are from (3.10) and (3.11), respectively. One can observe that as the number of antennas at each macro BS increases, more users are likely to associate with macro cells. This is because that the massive MIMO-aided macro cells are capable of providing larger array gain, which in turn enhance the average received power for the connected users. This observation is consistent with Remark 3.1. Another observation is that increasing the bias factor can encourage more users to connect to the small cells, which is an efficient method to extend the coverage of small cells or control loading balance among each tier of HetNets.

Figure 3.3 plots the coverage probability of a typical user associated with the k-tier NOMA enhanced small cells versus bias factor. The solid curves representing the analytical results of NOMA are from (3.21). One can observe that the coverage probability decreases as bias factor increases, which means that the unbiased user association outperforms the biased one, i.e., when $B_2 = 1$, the scenario becomes unbiased user association. This is because by invoking biased user association, users cannot be always associated with the BS which provides the highest received power. But the biased user association is capable of offering more flexibility for users as well as the whole networks, especially for the case that cells are fully overload. We also demonstrate that NOMA has superior behavior over OMA scheme.[2]

---

[2]The OMA benchmark adopted in this treatise is that by dividing the two users in equal time/frequency slots.

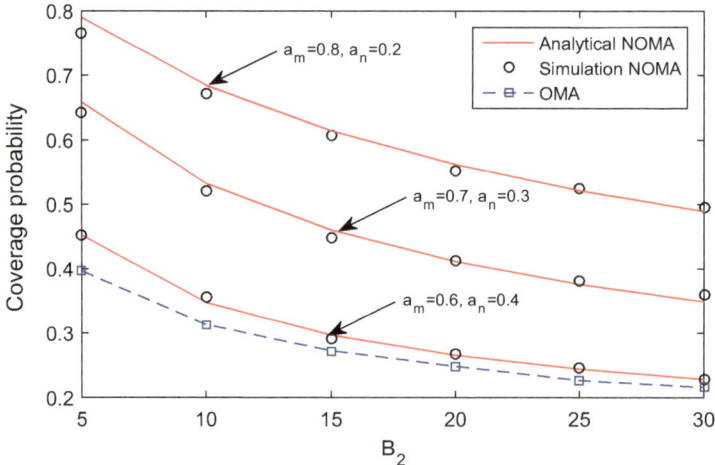

Fig. 3.3 Coverage probability comparison of NOMA- and OMA-based small cells. $K = 2$, $M = 200$, $N = 15$, $\lambda_2 = 20 \times \lambda_1$, $R_t = R_c = 1$ BPCU, $r_k = 10$ m, $P_1 = 40$ dBm, and $P_2 = 20$ dBm

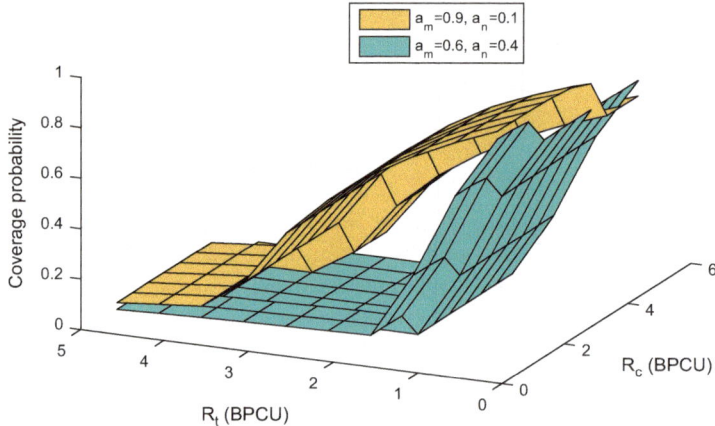

Fig. 3.4 Successful probability of typical user versus targeted rates of R_t and R_c, with $K = 2$, $M = 200$, $N = 15$, $\lambda_2 = 20 \times \lambda_1$, $r_k = 15$ m, $B_2 = 5$, $P_1 = 40$ dBm, and $P_2 = 20$ dBm

Figure 3.4 plots the coverage probability of a typical user associated with the k-tier NOMA enhanced small cells versus both R_t and R_c. We observe that there is a cross between these two plotted surfaces, which means that there exists an optimal power sharing allocation scheme for the given targeted rate. In contrast, for fixed power sharing coefficients, e.g., $a_m = 0.9, a_n = 0.1$, there also exist optimal targeted rates of two users for coverage probability. This figure also illustrates that for inappropriate power and targeted rate selection, the coverage probability is always zero, which also verifies our obtained insights in Remark 3.2.

3.1.5.2   Spectrum Efficiency

Figure 3.5 plots the spectrum efficiency of small cells with NOMA and OMA versus bias factor, B_2, with different transmit power of small cell BSs, P_2. The curves representing the performance of NOMA enhanced small cells are from (3.30). The performance of conventional OMA-based small cells is illustrated as a benchmark to demonstrate the effectiveness of our proposed framework. We observe that the spectrum efficiency of small cells decreases as the bias factor increases. This behavior can be explained as follows: larger bias factor associates more macro users with low SINR to small cells, which in turn degrades the spectrum efficiency of small cells. It is also worth noting that the performance of NOMA enhanced small cells outperforms the conventional OMA-based small cells, which in turn enhances the spectrum efficiency of the whole HetNets.

Figure 3.6 plots the spectrum efficiency of the proposed whole HetNets versus bias factor, B_2, with different transmit power, P_1. The curves representing the spectrum efficiency of small cells, macro cells, and HetNets are from (3.33). We can observe that macro cells can achieve higher spectrum efficiency compared to small cells. This is attributed to the fact that macro BSs are able to serve multiple users simultaneously with offering promising array gains to each user, which has been analytically demonstrated in Remark 3.3. It is also noted that the spectrum efficiency of macro cells improves as bias factor increases. The reason is again that when more low SINR macro cell users are associated with small cells, the spectrum efficiency of macro cells can be enhanced.

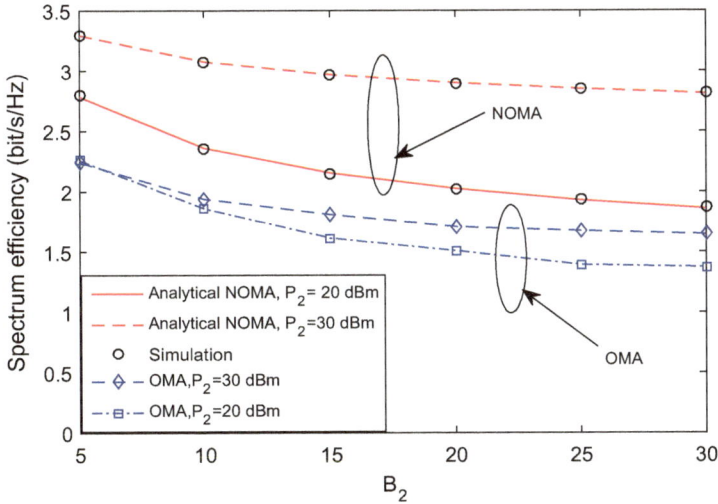

Fig. 3.5 Spectrum efficiency comparison of NOMA- and OMA-based small cells. $K = 2$, $M = 200$, $N = 15$, $r_k = 50$ m, $a_m = 0.6$, $a_n = 0.4$, $\lambda_2 = 20 \times \lambda_1$, and $P_1 = 40$ dBm

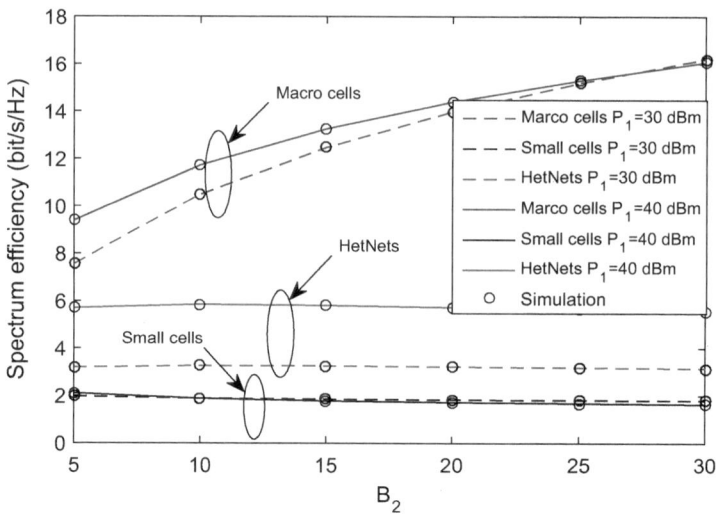

Fig. 3.6 Spectrum efficiency of the proposed framework. $r_k = 50$ m, $a_m = 0.6$, $a_n = 0.4$, $K = 2$, $M = 50$, $N = 5$, $P_2 = 20$ dBm, and $\lambda_2 = 100 \times \lambda_1$

3.1.5.3 Energy Efficiency

Figure 3.7 plots the energy efficiency of the proposed whole HetNets versus bias factor, B_2, with different transmit antenna of macro cell BSs, M. Several observations are as follows: (1) One observation is that the energy efficiency of the macro cells decreases as the number of antenna increases. Enlarging the number of antenna at the macro BSs is capable of offering a larger array gain, which in turn enhances the spectrum efficiency. Such operations also bring significant power consumption from the baseband signal processing of massive MIMO, which results in decreased energy efficiency. (2) Another observation is that NOMA enhanced small cells can achieve higher energy efficiency than the massive MIMO-aided macro cells. It means that from the perspective of energy consumption, densely deploying BSs in NOMA enhanced small cell is a more effective approach. (3) It is also worth noting that the number of antennas at the macro cell BSs almost has no effect on the energy efficiency of the NOMA enhanced small cells. (4) It also demonstrates that NOMA enhanced small cells have superior behavior than conventional OMA-based small cells in terms of energy efficiency. Such observations above demonstrate the benefits of the proposed NOMA enhanced hybrid HetNets and provide insightful guidelines for designing the practical large-scale networks.

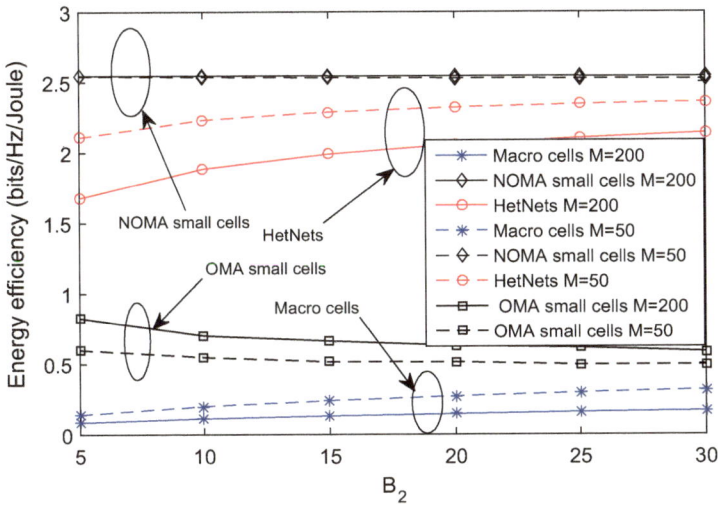

Fig. 3.7 Energy efficiency of the proposed framework. $K = 2$, $r_k = 10$ m, $a_m = 0.6$, $a_n = 0.4$, $N = 15$, $P_1 = 30$ dBm, $P_2 = 20$ dBm, and $\lambda_2 = 20 \times \lambda_1$

3.2   NOMA in Cognitive Radio Networks

The 2010s have witnessed the rapidly increasing penetration of mobile devices (e.g., smart phones, tablets, and laptops) all over the world, which gave rise to increasing demand for spectral resources. As reported by the Federal Communications Commission (FCC), there are significant temporal and spatial variations in the exploitation of the allocated spectrum. Given this fact, the CR concept inspired the community to mitigate the spectrum scarcity problem. The basic concept of CR is that at a certain time of the day or in a geographic region, the unlicensed secondary users (SUs) are allowed to opportunistically access the licensed spectrum of primary users (PUs). These CR techniques may be categorized into the interweave, overlay, and underlay paradigms:

- **Interweave**: The interweave CR can be regarded as an interference avoidance paradigm, where the SUs are required to sense the temporary slivers of the space-frequency domain of PUs before they access the channels (Qin et al. 2016a,b, 2018a). The concurrent transmission of SUs and PUs is not allowed under the interweave paradigm.
- **Overlay**: The overlay paradigm essentially constitutes an interference mitigation technique. With the aid of the classic dirty paper encoding technique, overlay CR ensures that a cognitive user becomes capable of transmitting simultaneously with a noncognitive PU (Goldsmith et al. 2009). Additionally, SUs are capable of forwarding the information of PUs to the PU receivers, while superimposing their own signals as a reward for their relaying services.

- **Underlay**: The underlay CR operates like an intelligent interference control paradigm, where the SUs are permitted to access the spectrum allocated to PUs as long as the interference power constraint at the PUs is satisfied.

One of the core challenges in both CR and NOMA networks is the interference management, while improving the bandwidth efficiency. Hence it is natural to link them for achieving an improved bandwidth efficiency. The application of NOMA in large-scale underlay CR networks has been investigated by using the stochastic geometry model (Liu et al. 2016d). The diversity order of the NOMA users was characterized analytically in two scenarios. The classic OMA-based underlay CR was also used as a benchmark to show the benefits of the proposed CR-NOMA scheme. Ding et al. (2016c) has proposed a novel power allocation (PA) policy for NOMA, namely the CR-inspired NOMA PA, which constitutes a beneficial amalgam of NOMA and underlay CR.

To the best of our knowledge, CR-NOMA studies only exist in the context of the underlay CR paradigm. Hence both the interweave and overlay CR paradigms have to be investigated in NOMA networks. It is worth pointing out that a significant research challenge of NOMA is to dynamically cluster/pair the NOMA users first, followed by dynamically allocating the clusters/pairs to different orthogonal sub-channels. In the context of the interweave paradigm, intelligent sensing has to be applied first, followed by user clustering/pairing of NOMA users, depending on the specific channel conditions sensed.

3.3 NOMA with MIMO

Multiple-antenna techniques are of significant importance, since they offer the extra dimension of the spatial domain, for further performance improvements. The application of multiple-antenna techniques in NOMA has attached substantial interest both from academia (Ding et al. 2016b,d; Hanif et al. 2016; Choi 2016; Kim et al. 2013; Qureshi et al. 2016; Liu et al. 2016a; Chen et al. 2016) and from industry (Higuchi and Kishiyama 2013; Higuchi and Benjebbour 2015; Benjebbour et al. 2013; Saito et al. 2013). The distinct NOMA features such as channel ordering and PA inevitably require special attention in the context of multiple antennas. More specifically, in contrast to the SISO-NOMA scenarios whose channels are all scalars, the channels of MIMO-NOMA scenarios are represented in form of matrices, which makes the power-based ordering of users rather challenging. As a consequence, conceiving an appropriate beamforming/precoding design is essential for multi-antenna-aided NOMA systems. NOMA relying on beamforming (BF) constitutes an efficient technique of improving the bandwidth efficiency by exploiting both the power domain and the angular domain. There are two popular MIMO-NOMA designs, namely the (1) Cluster-based (CB) MIMO-NOMA design; and the (2) Beamformer-based (BB) MIMO-NOMA design, which will be introduced in the following.

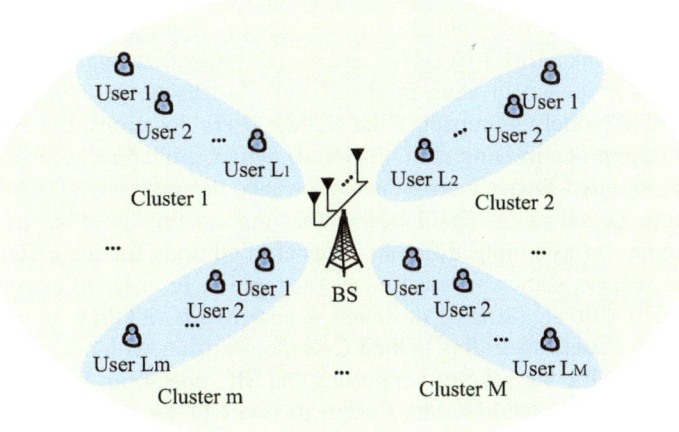

Fig. 3.8 Illustration of cluster-based MIMO-NOMA

3.3.1 Cluster-Based MIMO-NOMA

One of the popular NOMA designs is associated with the cluster-based structure, partitioning users into several different clusters. Explicitly, as shown in Fig. 3.8, the NOMA users are partitioned into M clusters and each cluster consists of L_m users, where $m \in \{1, 2, \ldots, M\}$. Then we design appropriate beams for the corresponding clusters. Upon applying effective transmit precoding and detector designs, it becomes possible to guarantee that the beam associated with a particular cluster is orthogonal to the channels of users in other clusters. Hence the inter-cluster interference can be efficiently suppressed. When considering each cluster in isolation, there is a difference among the users' channel conditions, hence we are faced again with the conventional NOMA scenarios. Thus, SIC can be readily invoked for mitigating the intra-cluster interference between users of the same cluster. Recently, many important research contributions investigated beamforming aided NOMA (Liu et al. 2016e).

Specifically, Choi (2015) proposed two-stage multicast zero-forcing (ZF)-based beamforming for downlink inter-group/cluster interference mitigation, where the total transmit power of each group/cluster was minimized during the second stage. Higuchi and Benjebbour (2015) utilized receive beamformers at the NOMA users and a transmit beamformer at the BS. Higuchi and Kishiyama (2013) then proposed a novel scheme, which combined open-loop random beamforming in conjunction with intra-beam SIC for downlink NOMA transmission. However, random beamforming fails to guarantee a constant QoS at the users' side. To overcome this limitation, in Ding et al. (2016a), Ding et al. proposed a TPC and detection scheme combination for a cluster-based downlink MIMO-NOMA scenario relying on fixed PA. By adopting this design, their MIMO-NOMA system can be decomposed into several independent single-input single-output (SISO) NOMA arrangements. Furthermore, in order to establish a more general framework considering both downlink and uplink MIMO-NOMA scenarios, the so-called signal alignment (SA) technique was proposed in Ding et al. (2016d). Stochastic geometry-based tools were invoked to model the impact of the NOMA users' locations. In contrast to the research contributions in Ding et al. (2016a,d), which are inter-cluster interference free design, an inter-cluster interference allowance design for CB MIMO-NOMA was proposed in Ali et al. (2017). Note that the existing NOMA designs have routinely relied on assuming different channel conditions for the different users, which is however a somewhat restrictive assumption. In order to circumvent this restriction, Ding et al. (2016b) designed a new MIMO-NOMA scheme, which distinguishes the users according to their QoS requirements with particular attention on IoT scenarios for the sake of supporting the SIC operation. Furthermore, they compared this new MIMO-NOMA design to two NOMA schemes, which order users according to the prevalent channel conditions. More particularly, the ZF-NOMA scheme of Ding et al. (2016a) and the SA-NOMA scheme (Ding et al. 2016d) were used as benchmarks in Ding et al. (2016b). Figure 3.9 illustrates the outage probability defined as the probability of erroneously detecting the message intended for User m in the i-th data stream, $i = 1, 2, 3$ at User n, where the QR decomposition is used to augmenting the differences between the users' effective channel conditions according to the associated QoS requirements. As shown in Fig. 3.9, the QR-based MIMO-NOMA scheme is capable of outperforming both ZF-NOMA and SA-NOMA[3] as well as MIMO-OMA,[4] since it exploits the heterogeneous QoS requirements of different users and applications. In Liu et al. (2016e), the fairness issues of the MIMO-NOMA scenario were addressed by applying appropriate user allocation algorithms among the clusters and dynamic PA algorithms within each cluster.

3.3.2 Beamformer-Based MIMO-NOMA

Another technique of implementing MIMO-NOMA is to assign different beams to different users, as shown in Fig. 3.10. By doing so, the QoS can be satisfied by calculating the beamformer-weights in a predefined order, commencing with the most demanding QoS requirement. By adopting this approach, several contributions have been made in terms of MIMO-NOMA. Considering the illustration of Fig. 3.10 as an example, User 1 to User N occupy the same RB, similar to user $(N+1)$ to user $(N+M)$. Again, within the same RB we may employ SIC at each user, according to

---

[3]Note that when $M = N$, ZF-NOMA achieves the same performance as SA-NOMA (Ding et al. 2016b).

[4]Figure 3.9 is focused on the performance of User n, since the QoS requirements have been guaranteed with the aid of appropriate PA (Ding et al. 2016b).

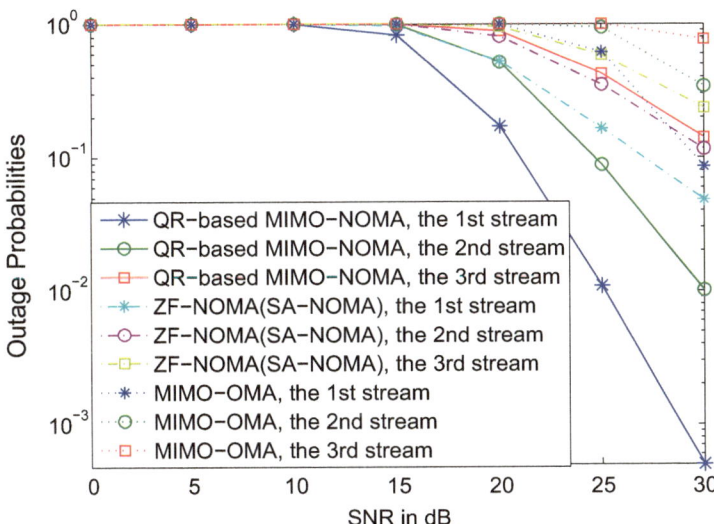

Fig. 3.9 Comparison of MIMO systems with various multiple-access techniques using a two-user case. $M = N = 3$, where M is the number of antennas at transmitters and N is the number of antennas at receivers. As such, each user is to receive three data streams from the BS. $R_m = 1.2$ bit per channel use (BPCU) and $R_n = 5$ BPCU are the targeted data rates of User m and User n, respectively

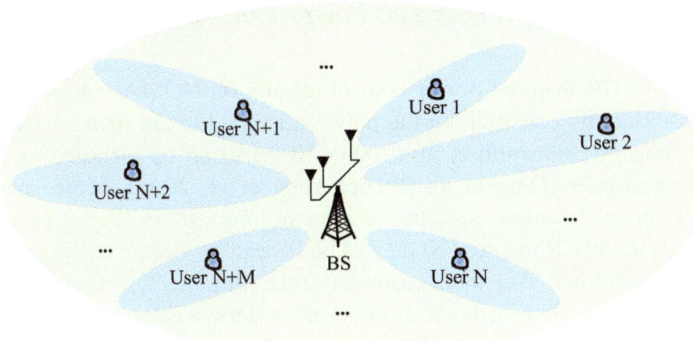

Fig. 3.10 Illustration of beamformer-based MIMO-NOMA

the particular ordering of the different users' received signal power. Sun et al. (2015) first investigated the power optimization problem constructed for maximizing the ergodic capacity and then showed that their proposed MIMO-NOMA schemes are capable of achieving significantly better performance than OMA. In an effort to reduce the decoding complexity imposed at the users, a layered transmission-based MIMO-NOMA scheme was proposed by Choi (2016), who also investigated the associated PA problem. It was demonstrated that upon invoking this layered transmission scheme, the achievable sum rate increases linearly with the number of antennas.

3.3.3 Massive-MIMO-NOMA

Massive MIMO may be considered as one of the key technologies Andrews et al. (2014) in 5G systems as a benefit of improving both the received SNR and the bandwidth efficiency. It was shown in Larsson et al. (2014) that massive MIMO is capable of substantially increasing both the capacity and the energy efficiency. These compelling benefits of massive MIMO sparked off the interest of researchers also in the context of NOMA. In Ding and Poor (2016), Ding and Poor conceived a two-stage TPC design for implementing massive-MIMO-NOMA. More particularly, a beamformer was adopted for serving to a cluster of angularly similar users and then they decomposed the MIMO-NOMA channels into a number of SISO-NOMA channels within the same cluster. A one-bit CSI feedback scheme was proposed for maintaining a low feedback overhead and a low implementation complexity.

3.3.4 Cognitive Radio Inspired Power Control

The objective of CR inspired power control relying on NOMA is to guarantee the QoS of weak users by constraining the power allocated to the strong user. Inspired by the CR concept Goldsmith et al. (2009), NOMA can be regarded as a special case of CR networks (Ding et al. 2016c; Yang et al. 2016). More specifically, still considering a downlink scenario supporting two users, Fig. 3.11 compares conventional CR and CR inspired NOMA. The BS can be viewed as the combination of a primary transmitter (PT) and a secondary transmitter (ST), which transmits the superimposed signals. The strong user (User n) and the weak user (User m) can be regarded as a secondary receiver (SR) and a primary receiver (PR), respectively. By doing so, the strong User n becomes capable of accessing the spectrum occupied by the weak User m under predetermined interference constraints, which is the key feature of the classic underlay CR. The concept of CR-inspired PA in NOMA was proposed by Ding et al. (2016c), who investigated the PA of user-pairing-based NOMA systems.

The key advantages of cognitive PA are summarized as follows:

- **Guaranteed QoS**: by applying cognitive PA, the QoS requirements of the weak user are guaranteed, which is especially vital in real-time safety-critical applications.

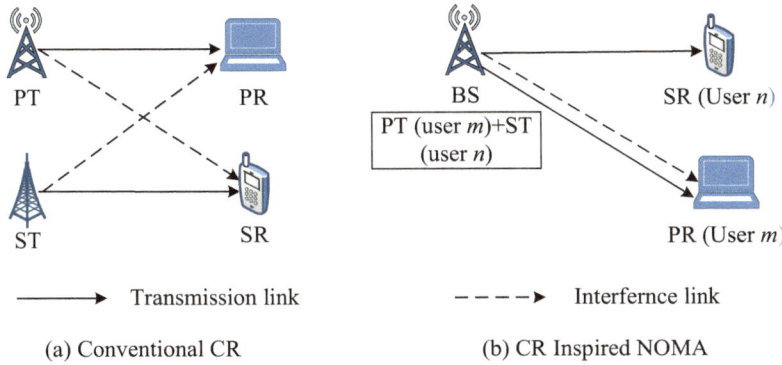

(a) Conventional CR   (b) CR Inspired NOMA

Fig. 3.11 Comparison of convectional CR (**a**) and CR inspired NOMA (**b**)

- **Fairness/throughput tradeoff**: cognitive PA is capable of striking a beneficial tradeoff between the overall system throughput and the individual user fairness, where the targeted data rate of the weak user has to be satisfied by appropriate PA.
- **High flexibility**: cognitive PA offers a high degree of freedom for the BS to explore the opportunistic support of the strong user.
- **Low complexity**: compared to the optimal PA approach, cognitive PA imposes a lower complexity during PA. This becomes particularly useful when the channel ordering and PA constraints are not convex and hence finding an appropriate PA scheme becomes a challenge, especially in multiple-antenna-aided NOMA scenarios.

Motivated by its advantages mentioned above, the cognitive PA policy was invoked for characterizing MIMO-NOMA systems. More particularly, in addition to investigating the convectional downlink cognitive PA conceived for MIMO-NOMA scenarios, the authors of Ding et al. (2016d) also designed a more sophisticated CR NOMA PA scheme for uplink MIMO-NOMA scenarios. In Ding et al. (2016b), in an effort to find a PA strategy suitable for SU-MIMO IoT scenarios, a cognitive PA policy was designed for ensuring that SIC may indeed be carried out at the strong user.

3.3.5 NOMA-Based Device-to-Device Communications

Due to the recent rapid increase in the demand for local area services under the umbrella of cellular networks, an emerging technique, namely device-to-device (D2D) communication, may be invoked for supporting direct communications among devices without the assistance of cellular BSs. The main advantages of integrating D2D communications into cellular networks are: (1) low-power support of proximity services for improving the energy efficiency; (2) reusing the frequency of the over-sailing cellular networks in an effort to increase the bandwidth efficiency; and (3) the potential to facilitate new types of peer-to-peer (P2P) services (Ma et al. 2015).

Note that one of the common features of both D2D and NOMA is that of enhancing the bandwidth efficiency by managing the interference among users within each RB. Motivated by this, it is desirable to invoke intelligent joint interference management approaches for fully exploiting the potential benefits of both D2D and NOMA. In Zhao et al. (2016), a novel NOMA-based D2D communication scheme has been designed, where several D2D groups were permitted to share the same RB with the cellular users. In contrast to the conventional D2D pair's transmission, the novel "D2D group" concept was introduced, where a D2D transmitter was able to simultaneously communicate with multiple D2D receivers with the aid of NOMA. It was demonstrated that the proposed NOMA-based D2D scheme is capable of delivering higher throughput than conventional D2D communications.

3.4 Summary

This chapter discusses the compatibility of NOMA when it is applied to the HetNets, MIMO, and CRNs techniques.

References

Adhikary, A., Dhillon, H. S., & Caire, G. (2015). Massive-MIMO meets HetNet: Interference coordination through spatial blanking. *IEEE Journal on Selected Areas in Communications, 33*, 1171–1186.

Ali, S., Hossain, E., & Kim, D. I. (2017). Non-orthogonal multiple access (NOMA) for downlink multiuser MIMO systems: User clustering, beamforming, and power allocation. *IEEE Access, 5*, 565–577.

Andrews, J. G., Buzzi, S., Choi, W., Hanly, S. V., Lozano, A., Soong, A. C., et al. (2014). What will 5G be? *IEEE Journal on Selected Areas in Communications, 32*, 1065–1082.

Benjebbour, A., Saito, Y., Kishiyama, Y., Li, A., Harada, A., & Nakamura, T. (2013). Concept and practical considerations of non-orthogonal multiple access (NOMA) for future radio access. In *Proceedings of IEEE Intelligent Signal Processing and Communications Systems (ISPACS)* (pp. 770–774).

Chen, Z., Ding, Z., Dai, X., & Karagiannidis, G. K. (2016). On the application of quasi-degradation to MISO-NOMA downlink. *IEEE Transactions on Signal Processing, 64*, 6174–6189.

Choi, J. (2015). Minimum power multicast beamforming with superposition coding for multiresolution broadcast and application to NOMA systems. *IEEE Transactions on Communications, 63*, 791–800.

Choi, J. (2016). On the power allocation for MIMO-NOMA systems with layered transmissions. *IEEE Transactions on Wireless Communications, 15*, 3226–3237.

Ding, Z., Adachi, F., & Poor, H. V. (2016a). The application of MIMO to non-orthogonal multiple access. *IEEE Transactions on Wireless Communications, 15*, 537–552.

Ding, Z., Dai, L., & Poor, H. V. (2016b). MIMO-NOMA design for small packet transmission in the internet of things. *IEEE Access, 4*, 1393–1405.

Ding, Z., Fan, P., & Poor, H. V. (2016c). Impact of user pairing on 5G non-orthogonal multiple access. *IEEE Transactions on Vehicular Technology, 65*, 6010–6023.

Ding, Z., & Poor, H. V. (2016). Design of massive-MIMO-NOMA with limited feedback. *IEEE Signal Processing Letters, 23*, 629–633.

Ding, Z., Schober, R., & Poor, H. V. (2016d). A general MIMO framework for NOMA downlink and uplink transmission based on signal alignment. *IEEE Transactions on Wireless Communications, 15*, 4438–4454.

Ding, Z., Yang, Z., Fan, P., & Poor, H. V. (2014). On the performance of non-orthogonal multiple access in 5G systems with randomly deployed users. *IEEE Signal Processing Letters, 21*, 1501–1505.

Goldsmith, A., Jafar, S. A., Maric, I., & Srinivasa, S. (2009). Breaking spectrum gridlock with cognitive radios: An information theoretic perspective. *Proceedings of the IEEE, 97*, 894–914.

Gradshteyn, I. S., & Ryzhik, I. M. (2000). *Table of integrals, series and products* (6th edn.). New York: Academic.

Hanif, M. F., Ding, Z., Ratnarajah, T., & Karagiannidis, G. K. (2016). A minorization-maximization method for optimizing sum rate in the downlink of non-orthogonal multiple access systems. *IEEE Transactions on Signal Processing, 64*, 76–88.

Higuchi, K., & Benjebbour, A. (2015). Non-orthogonal multiple access (NOMA) with successive interference cancellation for future radio access. *IEICE Transactions on Communications, 98*, 403–414.

Higuchi, K., & Kishiyama, Y. (2013). Non-orthogonal access with random beamforming and intra-beam SIC for cellular MIMO downlink. In *Proceedings of IEEE Vehicular Technology Conference (VTC Fall)* (pp. 1–5).

Hosseini, K., Yu, W., & Adve, R. S. (2014). Large-scale MIMO versus network MIMO for multicell interference mitigation. *IEEE Journal of Selected Topics in Signal Processing, 8*, 930–941.

Huh, H., Tulino, A. M., & Caire, G. (2012). Network MIMO with linear zero-forcing beamforming: Large system analysis, impact of channel estimation, and reduced-complexity scheduling. *IEEE Transactions on Information Theory, 58*, 2911–2934.

Jo, H.-S., Sang, Y. J., Xia, P., & Andrews, J. G. (2012). Heterogeneous cellular networks with flexible cell association: A comprehensive downlink SINR analysis. *IEEE Transactions on Wireless Communications, 11*, 3484–3495.

Kim, B., Lim, S., Kim, H., Suh, S., Kwun, J., Choi, S., et al. (2013). Non-orthogonal multiple access in a downlink multiuser beamforming system. In *Proceedings of Military Communications Conference (MILCOM)*, pp. 1278–1283.

Larsson, E., Edfors, O., Tufvesson, F., & Marzetta, T. (2014). Massive MIMO for next generation wireless systems. *IEEE Communications Magazine, 52*, 186–195.

Liu, L., Yuen, C., Guan, Y. L., & Li, Y. (2016a). Capacity-achieving iterative LMMSE detection for MIMO-NOMA systems. In *IEEE Proceedings of International Communication Conference (ICC), Kuala Lumpur, Malaysia*.

Liu, W., Jin, S., Wen, C. K., Matthaiou, M., & You, X. (2016b). A tractable approach to uplink spectral efficiency of two-tier massive MIMO cellular HetNets. *IEEE Communications Letters, 20*, 348–351.

Liu, Y., Ding, Z., Elkashlan, M., & Poor, H. V. (2016c). Cooperative non-orthogonal multiple access with simultaneous wireless information and power transfer. *IEEE Journal on Selected Areas in Communications, 34*(4), 938–953, April 2016.

Liu, Y., Ding, Z., Elkashlan, M., & Yuan, J. (2016d). Non-orthogonal multiple access in large-scale underlay cognitive radio networks. *IEEE Transactions on Vehicular Technology, 65*, 10152–10157.

Liu, Y., Elkashlan, M., Ding, Z., & Karagiannidis, G. K. (2016e). Fairness of user clustering in MIMO non-orthogonal multiple access systems. *IEEE Communications Letters, 20*, 1465–1468.

Ma, C., Wu, W., Cui, Y., & Wang, X. (2015). On the performance of successive interference cancellation in D2D-enabled cellular networks. In *Proceedings of IEEE International Conference on Computer Communication (INFOCOM), Kowloon, Hong Kong* (pp. 37–45)

Qin, Z., Fan, J., Liu, Y., Gao, Y., & Li, G. Y. (2018a). Sparse representation for wireless communications: A compressive sensing approach. *IEEE Signal Processing Magazine, 35,* 40–58.

Qin, Z., Gao, Y., & Parini, C. G. (2016a). Data-assisted low complexity compressive spectrum sensing on real-time signals under sub-Nyquist rate. *IEEE Transactions on Wireless Communications, 15,* 1174–1185.

Qin, Z., Gao, Y., Plumbley, M., & Parini, C. (2016b). Wideband spectrum sensing on real-time signals at sub-Nyquist sampling rates in single and cooperative multiple nodes. *IEEE Transactions on Signal Processing, 64,* 3106–3117.

Qin, Z., Liu, Y., Li, Y., & McCann, J. A. (2019). Performance analysis of clustered LoRa networks. In *IEEE Transactions on Vehicular Technology, 68*(8), 7616–7629, Aug. 2019.

Qin, Z., Yue, X., Liu, Y., Ding, Z., & Nallanathan, A. (2018b). User association and resource allocation in unified NOMA enabled heterogeneous ultra dense networks. *IEEE Communications Magazine, 56,* 86–92.

Qureshi, S., Kim, H., & Hassan, S. A. (2016). MIMO uplink NOMA with successive bandwidth division. In *Proceedings of IEEE Wireless Communication and Networking Conference, MILCOM, Doha*

Saito, Y., Kishiyama, Y., Benjebbour, A., Nakamura, T., Li, A., & Higuchi, K. (2013). Non-orthogonal multiple access (NOMA) for cellular future radio access. In *IEEE Proceedings of Vehicle Technology Conference (VTC), Dresden* (pp. 1–5).

Sun, Q., Han, S., I, C.-L., & Pan, Z. (2015). On the ergodic capacity of MIMO NOMA systems. *IEEE Wireless Communications Letters, 4,* 405–408.

Yang, Z., Ding, Z., Fan, P., & Al-Dhahir, N. (2016). A general power allocation scheme to guarantee quality of service in downlink and uplink NOMA systems. *IEEE Transactions on Wireless Communications, 15,* 7244–7257.

Ye, Q., Bursalioglu, O. Y., Papadopoulos, H. C., Caramanis, C., & Andrews, J. G. (2015). User association and interference management in massive MIMO HetNets. arXiv preprint arXiv:1509.07594.

Zhao, J., Liu, Y., Chai, K. K., Chen, Y., Elkashlan, M., & Alonso-Zarate, J. (2016). NOMA-based D2D communications towards 5G. In *IEEE Proceedings of Global Communication Conference (GLOBECOM), Washington* (pp. 1–6).

Chapter 4
Sustainability of NOMA

In this chapter, the sustainability of NOMA will be discussed by talking about cooperative NOMA and wireless powered NOMA networks. Particularly, the average performance of wireless powered NOMA will be presented in detail as an example.

4.1 Cooperative NOMA Networks

4.1.1 Cooperative NOMA

The key idea behind cooperative NOMA is to rely on strong NOMA users acting as DF relays to assist weak NOMA users. Still considering the two-user downlink transmission of Fig. 2.2b as our example, cooperative NOMA requires two time slots for its transmission. The first slot is for the direct transmission phase, which is the same as the noncooperative NOMA of Fig. 2.2b indicated by solid lines. During the second time slot, which is the cooperative phase, User n will forward the decoded message to User m by invoking the DF relaying protocol of Fig. 2.2b indicated by the dashed line. Proposed by Ding et al. (2015), this novel concept intrigued researchers, since cooperative NOMA fully benefits from SIC and DF decoding. In Ding et al. (2015), a general downlink NOMA transmission scenario was considered, where the BS supported M users with the aid of cooperative NOMA protocols relying on M slots. In an effort to seek a more efficient cooperative NOMA protocol, Liu et al. (2016) proposed a new EH assisted cooperative NOMA scheme. A sophisticated stochastic geometry-based model was invoked for evaluating the system's performance and user pairing was adopted for reducing the implementation complexity. Compared to conventional NOMA, the key advantages of cooperative NOMA transmissions can be summarized as follows:

© The Author(s), under exclusive license to Springer Nature Switzerland AG 2020
Y. Liu et al., *Non-Orthogonal Multiple Access for Massive Connectivity*,
SpringerBriefs in Computer Science, https://doi.org/10.1007/978-3-030-30975-6_4

- **Low system redundance**: Again, upon applying SIC techniques in NOMA, the message of the weak user has already been decoded at the strong user, hence it is natural to consider the employment of the DF protocol for weak signal. Explicitly, the weak signal can be remodulated and retransmitted from a position closer to the destination.
- **Better fairness**: A beneficial feature of cooperative NOMA is that the reliability of the weak user is significantly improved. As a consequence, the fairness of NOMA transmission can be improved (Timotheou and Krikidis 2015), particularly in the scenarios when the weak user is at the edge of the cell illustrated by the BS.
- **Higher diversity gain**: Cooperative NOMA is capable of achieving an improved diversity gain for the weak NOMA user, which is an effective technique of overcoming multi-path fading. It was analytically demonstrated (Ding et al. 2015) that the diversity gains of the weak users in cooperative NOMA are the same as those of the conventional cooperative networks, even for using EH relays (Liu et al. 2016).

Figure 4.1 illustrates the superior performance of cooperative NOMA over noncooperative NOMA as well as over OMA in terms of its outage probability. It is noted that cooperative NOMA achieves a higher diversity gain than noncooperative NOMA and OMA, which demonstrates the effectiveness of cooperative NOMA, as aforementioned. Note that cooperative NOMA constitutes one of many techniques of improving the transmission reliability of NOMA networks, especially for the weak NOMA users who have poor channel conditions. There are also other techniques for enhancing the performance of NOMA networks, which will be detailed

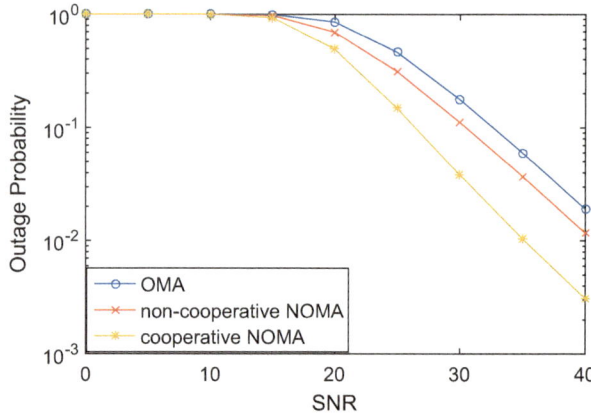

Fig. 4.1 Performance comparison between cooperative NOMA and noncooperative NOMA. The BS is located at (0, 0), User m is located at (4, 0), User n is located at (1, 0), the path loss exponent is 3, and the power allocation coefficients for User m and User n are (0.8, 0.2). The targeted data rate is 1 bits per channel use (BPCU)

in the following subsection. It is also worth pointing out that the performance of cooperative NOMA is related to the error propagation issue for SIC.

4.1.2 NOMA in Cooperative Transmission-Based Networks

Cooperative communication and NOMA techniques mutually support each other, hence the joint action of both techniques further improves the cooperative network performance (Men and Ge 2015a,b; Choi 2014; Kim and Lee 2015). Relaying and CoMP transmission are capable of improving cooperative networks.

4.1.2.1 Relay-Aided NOMA Transmission

Relaying has recently attracted considerable attention as a benefit of its network coverage extension (Laneman et al. 2004). In this context, Men and Ge investigated the outage performance of the single-antenna AF relay-aided NOMA downlink (Men and Ge 2015b). By contrast, multiple antennas were used by Men and Ge (2015a). The potential gains of NOMA over OMA were quantified in both scenarios. As a further development, a DF-based two-stage relay selection protocol was proposed in Ding et al. (2016), which relied on maximizing the diversity gain and minimizing the outage probability. Duan et al. (2016) proposed a novel two-stage PA scheme for a dual-hop relay-aided NOMA system, where the destination jointly decoded the information received both from the source and from the relay by applying the classic MRC technique. Kim and Lee (2015) considered a coordinated direct and relay-aided transmission scheme, where the BS simultaneously transmitted both to a nearby user and to a relay by invoking NOMA techniques during the first phase, while reaching a distant user with the aid of the relay.

4.1.2.2 Multi-Cell NOMA with Coordinated Multipoint Transmission

When considering multi-cell scenarios, the performance of the cell-edge users is of particular concern. This is particularly important for downlink NOMA, since the SIC operations are usually carried out for cell-center users rather than for cell-edge users. Hence the cell-edge users may not be well served. CoMP transmissions constitute an effective technique of improving the performance of cell-edge users. The key concept of CoMP in multi-cell NOMA is to enable multiple BSs to carry out coordinated beamforming or joint signal processing for the cell-edge users (Ali et al. 2017). There are several research contributions in the context of handling the inter-cell interference in NOMA networks (Choi 2014; Shin et al. 2017; Tian et al. 2016).

Choi (2014) incorporated NOMA into CoMP for the sake of attaining a bandwidth efficiency improvement. A new coordinated superposition coding scheme relying on Alamouti's space-time code was proposed. As a further advance, Shin et al. (2017) investigated the performance of multi-cell MIMO-NOMA networks, applying coordinated beamforming for dealing with the inter-cell interference in order to enhance the cell-edge users' throughput. Tian et al. (2016) conceived an opportunistic NOMA scheme for CoMP systems and compared it to the conventional joint-transmission-based NOMA. Vien et al. (2015) proposed a NOMA-based PA policy for cloud radio access network (C-RAN) scenarios (Yang et al. 2019), which can also be regarded as a coordinated transmission scenario, where the BSs jointly form a cloud. Martin-Vega et al. (2017) proposed a novel NOMA-aided C-RAN network, associated with using stochastic geometry. It revealed that the proposed framework is capable of substantially enhancing the performance of cell-edge users.

4.2 Wireless Powered NOMA

In addition to improving spectral efficiency which is the motivation of NOMA, another key objective of future 5G networks is to maximize energy efficiency. Simultaneous wireless information and power transfer (SWIPT), which was initially proposed by Varshney (2008), has rekindled the interest of researchers to explore more energy efficient networks by Qin et al. (2017). It was assumed that both information and energy could be extracted from the same radio frequency signals at the same time, which does not hold in practice. Motivated by this issue, two practical receiver architectures, namely time switching (TS) receiver and power splitting (PS) receiver, were proposed in a multi-input and multi-output (MIMO) system.

One important advantage of the NOMA concept is that it can squeeze a user with better channel conditions into the bandwidth channel which has been occupied by a user with poor channel conditions. For example, consider a practical scenario, in which there are two groups of users: (1) near users, which are close to the base station (BS) and have better channel conditions; and (2) far users, which are close to the edge of the cell controlled by the BS and therefore have poor channel conditions. While the spectral efficiency of NOMA is superior compared to orthogonal MA, the fact that the near users coexist with the far users brings performance degradation at the far users. It is natural to consider the use of the near users as DF relays to transmit information to the far users. In order to extend the lifetime of NOMA users, we consider that the near users is capable of harvesting energy from their received RF signals. To improve the reliability of the far NOMA users without draining the near users' batteries, we consider the application of SWIPT to NOMA, where SWIPT is performed at the near NOMA users. Therefore, the aforementioned two communication concepts, cooperative NOMA and SWIPT, can be naturally linked together.

Fig. 4.2 An illustration of SWIPT NOMA considering a BS (blue circle). The spatial distributions of the near users (yellow circles) and the far users (green circles) follow homogeneous PPPs

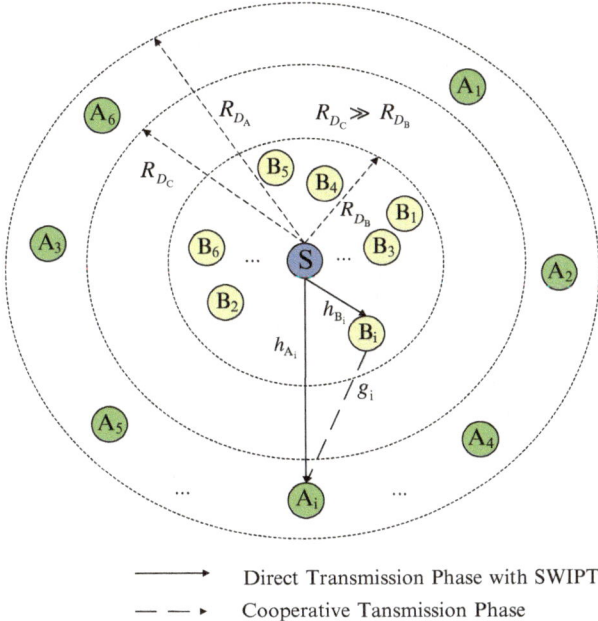

—————▶ Direct Transmission Phase with SWIPT

— — ▶ Cooperative Tansmission Phase

4.2.1 Network Model

We consider a network with a single source S (i.e., the base station (BS)) and two groups of randomly deployed users $\{A_i\}$ and $\{B_i\}$. We assume that the users in group $\{B_i\}$ are deployed within disc D_B with radius R_{D_B}. The far users $\{A_i\}$ are deployed within ring D_A with radius R_{D_C} and R_{D_A} (assuming $R_{D_C} \gg R_{D_B}$), as shown in Fig. 4.2. Note that the BS is located at the origin of both the disc D_B and the ring D_A. The locations of the near and far users are modeled as homogeneous PPPs Φ_κ ($\kappa \in \{A, B\}$) with density λ_{Φ_κ}. Here the near users are uniformly distributed within the disc and the far users are uniformly distributed within the ring. The number of users in R_{D_κ}, denoted by N_κ, follows a Poisson distribution as $\Pr(N_\kappa = k) = (\mu_\kappa^k/k!)e^{-\mu_\kappa}$, where μ_κ is the mean measure, i.e., $\mu_A = \pi\left(R_{D_A}^2 - R_{D_C}^2\right)\lambda_{\Phi_A}$ and $\mu_B = \pi R_{D_B}^2 \lambda_{\Phi_B}$. All channels are assumed to be quasi-static Rayleigh fading, where the channel coefficients are constant for each transmission block but vary independently between different blocks. In the proposed network, we consider that the users in $\{B_i\}$ are energy harvesting relays that harvest energy from the BS and forward the information to $\{A_i\}$ using the harvested energy as their transmit power. The DF strategy is applied at $\{B_i\}$ and the cooperative NOMA system consists of two phases, detailed in the following. In this work, without loss of generality, it is assumed that the two phases have the same transmission periods, the same as in Nasir et al. (2013), Ding et al. (2014a), and Ding and Poor (2013). It is worth pointing out that dynamic time allocation for the two phases can further improve the performance of the proposed cooperative NOMA scheme, which is beyond the scope of the paper.

4.2.1.1 Phase 1: Direct Transmission

Prior to transmission, the two users denoted by A_i and B_i are selected to perform NOMA, where the selection criterion will be discussed in the next section. During the first phase, the BS sends two messages $p_{i1}x_{i1} + p_{i2}x_{i2}$ to two selected users A_i and B_i based on NOMA (Ding et al. 2014b), where p_{i1} and p_{i2} are the power allocation coefficients and x_{i1} and x_{i2} are the messages of A_i and B_i, respectively. The observation at A_i is given by

$$y_{A_i,1} = \sqrt{P_S} \sum_{k \in \{1,2\}} p_{ik} x_{ik} \frac{h_{A_i}}{\sqrt{1 + d_{A_i}^{\alpha}}} + n_{A_i,1}, \qquad (4.1)$$

where P_S is the transmit power at the BS, h_{A_i} models the small-scale Rayleigh fading from the BS to A_i with $h_{A_i} \sim \mathcal{CN}(0, 1)$, $n_{A_i,1}$ is the additive Gaussian white noise (AWGN) at A_i with variance $\sigma_{A_i}^2$, d_{A_i} is the distance between BS and A_i, and α is the path loss exponent.

Without loss of generality, we assume that $|p_{i1}|^2 > |p_{i2}|^2$ with $|p_{i1}|^2 + |p_{i2}|^2 = 1$. The received signal to interference and noise ratio (SINR) at A_i to detect x_{i1} is given by

$$\gamma_{S,A_i}^{x_{i1}} = \frac{\rho |h_{A_i}|^2 |p_{i1}|^2}{\rho |p_{i2}|^2 |h_{A_i}|^2 + 1 + d_{A_i}^{\alpha}}, \qquad (4.2)$$

where $\rho = \frac{P_S}{\sigma^2}$ is the transmit signal to noise radio (SNR) (assuming $\sigma_{A_i}^2 = \sigma_{B_i}^2 = \sigma^2$).

We consider that the near users have the rechargeable storage ability (Nasir et al. 2013) and power splitting (Zhang and Ho 2013) is applied to perform SWIPT. From the implementation point of view, this rechargeable storage unit can be a supercapacitor or a short-term high-efficiency battery (Krikidis et al. 2014). The power splitting approach is applied as explained in the following: the observation at B_i is divided into two parts. One part is used for information decoding by directing the observation flow to the detection circuit and the remaining part is used for energy harvesting to power B_i for helping A_i. Thus,

$$y_{B_i,1} = \sqrt{P_S} \sum_{k \in \{1,2\}} p_{ik} x_{ik} \frac{\sqrt{1 - \beta_i} h_{B_i}}{\sqrt{1 + d_{B_i}^{\alpha}}} + n_{B_i,1}, \qquad (4.3)$$

where β_i is the power splitting coefficient which is detailed in (4.7), h_{B_i} models the small-scale Rayleigh fading from BS to B_i with $h_{B_i} \sim \mathcal{CN}(0, 1)$, n_{B_i} is AWGN at $n_{B_i,1}$ with variance $\sigma_{B_i}^2$, and d_{B_i} is the distance between BS and B_i. We use the bounded path loss model to ensure that the path loss is always larger than one even for small distances (Ding et al. 2014a).

Applying NOMA, the successive interference cancelation (SIC) (Cover and Thomas 2006) is carried out at B_i. Particularly, B_i first decodes the message of A_i, then subtracts this component from the received signal to detect its own information. Therefore, the received SINR at B_i to detect x_{i1} of A_i is given by

$$\gamma_{S,B_i}^{x_{i1}} = \frac{\rho |h_{B_i}|^2 |p_{i1}|^2 (1 - \beta_i)}{\rho |h_{B_i}|^2 |p_{i2}|^2 (1 - \beta_i) + 1 + d_{B_i}^\alpha}. \tag{4.4}$$

The received SNR at B_i to detect x_{i2} of B_i is given by

$$\gamma_{S,B_i}^{x_{i2}} = \frac{\rho |h_{B_i}|^2 |p_{i2}|^2 (1 - \beta_i)}{1 + d_{B_i}^\alpha}. \tag{4.5}$$

The power splitting coefficient β_i is used to determine the amount of harvested energy. Based on (4.4), the data rate supported by the channel from BS to B_i for decoding x_{i1} is given by

$$R_{x_{i1}} = \frac{1}{2} \log \left(1 + \frac{\rho |h_{B_i}|^2 |p_{i1}|^2 (1 - \beta_i)}{\rho |h_{B_i}|^2 |p_{i2}|^2 (1 - \beta_i) + 1 + d_{B_i}^\alpha} \right) !`\$ \tag{4.6}$$

We assume that the energy required to receive/process information is negligible compared to the energy required for information transmission (Nasir et al. 2013). In this work, we apply the dynamic power splitting protocol which means that the power splitting coefficient β_i is a variable and opportunistically tuned to support the relay transmission. Our aim is to first guarantee the detection of the message of the far NOMA user, A_i, at the near NOMA user B_i, then B_i can harvest the remaining energy. In this case, based on (4.6), in order to ensure that B_i can successfully decode the information of, A_i, at a given target rate, i.e., $R_1 = R_{x_{i1}}$. Therefore, the power splitting coefficient is set as follows:

$$\beta_i = \max \left\{ 0, 1 - \frac{\tau_1 \left(1 + d_{B_i}^\alpha \right)}{\rho \left(|p_{i1}|^2 - \tau_1 |p_{i2}|^2 \right) |h_{B_i}|^2} \right\}, \tag{4.7}$$

where $\tau_1 = 2^{2R_1} - 1$. Here $\beta_i = 0$ means that all the energy is used for information decoding and no energy is remaining for energy harvesting.

Based on (4.3), the energy harvested at B_i is given by

$$E_{B_i} = \frac{T \eta P_S \beta_i \left| h_{B_i} \right|^2}{2 \left(1 + d_{B_i}^\alpha \right)}, \tag{4.8}$$

where T is the time period for the whole transmission including the direct transmission phase and the cooperative transmission phase, and η is the energy harvesting coefficient. We assume that the two phases have the same transmission period; therefore, the transmit power at B_i is expressed as follows:

$$P_t = \frac{\eta P_S \beta_i \left| h_{B_i} \right|^2}{1 + d_{B_i}^\alpha}. \tag{4.9}$$

4.2.1.2   Phase 2: Cooperative Transmission

During this phase, B_i forwards x_{i1} to A_i by using the harvested energy during the direct transmission phase. In this case, A_i observes

$$y_{A_i,2} = \frac{\sqrt{P_t} x_{i1} g_i}{\sqrt{1 + d_{C_i}^\alpha}} + n_{A_i,2}, \tag{4.10}$$

where g_i models the small-scale Rayleigh fading from B_i to A_i with $g_i \sim \mathcal{CN}(0, 1)$, $n_{A_i,2}$ is AWGN at A_i with variance $\sigma_{A_i}^2$, $d_{C_i} = \sqrt{d_{A_i}^2 + d_{B_i}^2 - 2 d_{A_i} d_{B_i} \cos(\theta_i)}$ is the distance between B_i and A_i, and θ_i denotes the angle $\angle A_i S B_i$.

Based on (4.9) and (4.10), the received SNR for A_i to detect x_{i1} forwarded from B_i is given by

$$\gamma_{A_i,B_i}^{x_{i1}} = \frac{P_t \left| g_i \right|^2}{\left(1 + d_{C_i}^\alpha \right) \sigma^2} = \frac{\eta \rho \beta_i \left| h_{B_i} \right|^2 \left| g_i \right|^2}{\left(1 + d_{C_i}^\alpha \right) \left(1 + d_{B_i}^\alpha \right)}. \tag{4.11}$$

At the end of this phase, A_i combines the signals from BS and B_i using maximal-ratio combining (MRC). Combining the SNR of the direct transmission phase (4.2) and the SINR of the cooperative transmission phase (4.11), we obtain the received SINR at A_i as follows:

$$\gamma_{A_i,\mathrm{MRC}}^{x_{i1}} = \frac{\rho \left| h_{A_i} \right|^2 \left| p_{i1} \right|^2}{\rho \left| h_{A_i} \right|^2 \left| p_{i2} \right|^2 + 1 + d_{A_i}^\alpha} + \frac{\eta \rho \beta_i \left| h_{B_i} \right|^2 \left| g_i \right|^2}{\left(1 + d_{B_i}^\alpha \right) \left(1 + d_{C_i}^\alpha \right)}. \tag{4.12}$$

4.2.2   Non-orthogonal Multiple Access with User Selection

In this section, the performance of three user selection schemes is characterized in the following.

4.2.2.1   RNRF Selection Scheme

In this scheme, the BS randomly selects a near user B_i and a far user A_i. This selection scheme provides a fair opportunity for each user to access the source with the NOMA protocol. The advantage of this user selection scheme is that it does not require the knowledge of instantaneous channel state information (CSI). To make meaningful conclusions, in the rest of the paper, we only focus on $\beta_i > 0$ and the number of the near users and the far users both satisfy $N_B \geq 1, N_A \geq 1$.

In the NOMA protocol, an outage of B_i can occur due to two reasons. The first is when B_i cannot detect x_{i1}. The second is when B_i can detect x_{i1} but cannot detect x_{i2}. To guarantee that the NOMA protocol can be implemented, the condition $|p_{i1}|^2 - |p_{i2}|^2 \tau_1 > 0$ should be satisfied (Ding et al. 2014b). Based on this, the outage probability of B_i can be expressed as follows:

$$P_{B_i} = \Pr\left(\frac{\rho|h_{B_i}|^2|p_{i1}|^2}{\rho|h_{B_i}|^2|p_{i2}|^2 + 1 + d_{B_i}^\alpha} < \tau_1\right) \tag{4.13}$$

$$+ \Pr\left(\frac{\rho|h_{B_i}|^2|p_{i1}|^2}{\rho|h_{B_i}|^2|p_{i2}|^2 + 1 + d_{B_i}^\alpha} > \tau_1, \gamma_{S,B_i}^{x_{i2}} < \tau_2\right),$$

where $\tau_2 = 2^{2R_2} - 1$ with R_2 being the target rate at which B_i can detect x_{i2}.

The following theorem provides the outage probability of the near users in RNRF for an arbitrary choice of α.

Theorem 4.1 *Conditioned on PPP, the outage probability of the near users B_i can be approximated in closed-form as follows:*

$$P_{B_i} \approx \frac{1}{2} \sum_{n=1}^{N} \omega_N \sqrt{1 - \phi_n^2} \left(1 - e^{-c_n \varepsilon_{A_i}}\right) (\phi_n + 1), \tag{4.14}$$

if $\varepsilon_{A_i} \geq \varepsilon_{B_i}$, otherwise $P_{B_i} = 1$, where $\varepsilon_{A_i} = \frac{\tau_1}{\rho(|p_{i1}|^2 - |p_{i2}|^2 \tau_1)}$ and $\varepsilon_{B_i} = \frac{\tau_2}{\rho|p_{i2}|^2}$, N is a parameter to ensure a complexity-accuracy tradeoff, $c_n = 1 + \left(\frac{RD_B}{2}(\phi_n + 1)\right)^\alpha$, $\omega_N = \frac{\pi}{N}$, and $\phi_n = \cos\left(\frac{2n-1}{2N}\pi\right)$.

Corollary 4.1 *For the special case* $\alpha = 2$, *the outage probability of* B_i *can be obtained as exact expression as follows:*

$$P_{B_i}\big|_{\alpha=2} = 1 - \frac{e^{-\varepsilon_{A_i}}}{R_{D_B}^2 \varepsilon_{A_i}} + \frac{e^{-\left(1+R_{D_B}^2\right)\varepsilon_{A_i}}}{R_{D_B}^2 \varepsilon_{A_i}}, \tag{4.15}$$

if $\varepsilon_{A_i} \geq \varepsilon_{B_i}$, *otherwise* $P_{B_i}\big|_{\alpha=2} = 1$.

With the proposed cooperative SWIPT NOMA protocol, the outage experienced by A_i can occur due to two reasons. The first is when B_i can detect x_{i1} but the overall received SNR at A_i cannot support the targeted rate. The second is that neither A_i nor B_i can detect x_{i1}. Based on this, the outage probability can be expressed as follows:

$$P_{A_i} = \Pr\left(\gamma_{A_i,MRC}^{x_{i1}} < \tau_1, \gamma_{S,B_i}^{x_{i1}}\big|_{\beta_i=0} > \tau_1\right) + \Pr\left(\gamma_{S,A_i}^{x_{i1}} < \tau_1, \gamma_{S,B_i}^{x_{i1}}\big|_{\beta_i=0} < \tau_1\right). \tag{4.16}$$

The following theorem provides the outage probability of the far users in RNRF for an arbitrary choice of α.

Theorem 4.2 *Conditioned on PPP, assume* $R_{D_C} \gg R_{D_B}$, *the outage probability of* A_i *can be approximated in closed-form as follows:*

$$P_{A_i} \approx \zeta_1 \sum_{n=1}^{N} (\phi_n + 1) \sqrt{1 - \phi_n^2} c_n \sum_{k=1}^{K} \sqrt{1 - \psi_k^2} s_k \left(1 + s_k^\alpha\right)^2$$

$$\times \sum_{m=1}^{M} \sqrt{1 - \varphi_m^2} e^{-(1+s_k^\alpha)t_m} \chi_{t_m} \left(\ln \frac{\chi_{t_m}\left(1 + s_k^\alpha\right)}{\eta\rho} c_n + 2c_0\right)$$

$$+ a_1 \sum_{n=1}^{N} \sqrt{1 - \phi_n^2} c_n (\phi_n + 1) \sum_{k=1}^{K} \sqrt{1 - \psi_k^2}(1 + s_k^\alpha) s_k, \tag{4.17}$$

where M and K are parameters to ensure a complexity-accuracy tradeoff, $\zeta_1 = -\frac{\varepsilon_{A_i} R_{D_B} \omega_N \omega_K \omega_M}{8\left(R_{D_{A_i}} + R_{D_{C_i}}\right)\eta\rho}$, $\chi_{t_m} = \tau_1 - \frac{\rho t_m |p_{i1}|^2}{\rho t_m |p_{i2}|^2 + 1}$, $t_m = \frac{\varepsilon_{A_i}}{2}(\varphi_m + 1)$, $\omega_M = \frac{\pi}{M}$, $\varphi_m = \cos\left(\frac{2m-1}{2M}\pi\right)$, $s_k = \frac{R_{D_A} - R_{D_C}}{2}(\psi_k + 1) + R_{D_C}$, $\omega_K = \frac{\pi}{K}$, $\psi_k = \cos\left(\frac{2k-1}{2K}\pi\right)$, $c_0 = -\frac{\varphi(1)}{2} - \frac{\varphi(2)}{2}$, *and* $a_1 = \frac{\omega_K \omega_N \varepsilon_{A_1}^2}{2(R_{D_A} + R_{D_C})}$.

Corollary 4.2 *For the special case* $\alpha = 2$, *the outage probability of* A_i *can be simplified in closed-form as follows:*

$$P_{A_i}\big|_{\alpha=2} \approx \zeta_2 \sum_{k=1}^{K} \sqrt{1 - \psi_k^2} s_k \left(1 + s_k^2\right)^2 \sum_{m=1}^{M} \sqrt{1 - \varphi_m^2}$$

$$\times \chi_{t_m} e^{-\left(1+s_k^2\right)t_m} \left(\ln \frac{\chi_{t_m}\left(1+s_k^2\right)}{\eta\rho} c_n + b_0 \right)$$

$$+ \left(1 - \frac{e^{-\left(1+R_{DC}^2\right)\varepsilon_{A_i}}}{\varepsilon_{A_i}\left(R_{DA}^2 - R_{DC}^2\right)} + \frac{e^{-\left(1+R_{DA}^2\right)\varepsilon_{A_i}}}{\varepsilon_{A_i}\left(R_{DA}^2 - R_{DC}^2\right)} \right)$$

$$\times \left(1 - \frac{e^{-\varepsilon_{A_i}}}{R_{DB}^2\varepsilon_{A_i}} + \frac{e^{-\left(1+R_{DB}^2\right)\varepsilon_{A_i}}}{R_{DB}^2\varepsilon_{A_i}} \right), \tag{4.18}$$

where $\zeta_2 = -\dfrac{\omega_K\omega_M\varepsilon_{A_i}\left(R_{DB}^2+2\right)}{8\left(R_{DA}+R_{DC}\right)\eta\rho}$ and $b_0 = \dfrac{\left(1+R_{DB}^2\right)^2\ln\left(1+R_{DB}^2\right)}{2R_{DB}^2} + \left(R_{DB}^2+2\right)\left(c_0 - \frac{1}{4}\right)$.

To obtain further insights for the derived outage probability, we provide the diversity analysis of both the near and far uses of RNRF.

Near Users For the near users, based on the analytical results, we carry out high SNR approximations as follows. When $\varepsilon \to 0$, we can have

$$F_{Y_i}(\varepsilon) \approx \frac{1}{2}\sum_{n=1}^{N}\omega_N\sqrt{1 - \phi_n^2}c_n\varepsilon_{A_i}\left(\phi_n + 1\right). \tag{4.19}$$

The diversity gain is defined as follows:

$$d = -\lim_{\rho\to\infty}\frac{\log P(\rho)}{\log\rho}. \tag{4.20}$$

Substituting (4.19) into (4.20), we obtain that the diversity gain for the near users is one, which means that using NOMA with energy harvesting will not decrease the diversity gain.

Far Users For the far users, substituting (4.17) into (4.20), we obtain

$$d = -\lim_{\rho\to\infty}\frac{\log\left(-\frac{1}{\rho^2}\log\frac{1}{\rho}\right)}{\log\rho} = -\lim_{\rho\to\infty}\frac{\log\log\rho - \log\rho^2}{\log\rho} = 2. \tag{4.21}$$

As we can see from (4.21), the diversity gain of RNRF is two, which is the same as that of the conventional cooperative network (Laneman et al. 2004). This result indicates that using NOMA with an energy harvesting relay will not affect the diversity gain. In addition, we see that in high SNRs, the dominant factor for the outage probability is $\frac{1}{\rho^2} \ln \rho$. Therefore we conclude that the outage probability of using NOMA with SWIPT decays at a rate of $\frac{\ln SNR}{SNR^2}$. However, for a conventional cooperative system without energy harvesting, a faster decreasing rate of $\frac{1}{SNR^2}$ can be achieved.

4.2.2.2 NNNF Selection Scheme

In this subsection, we characterize the performance of NNNF, which exploits the users' CSI opportunistically. We first select a user within the disc D_B which has the shortest distance to the BS as the near NOMA user (denoted by B_{i*}). This is because the near users also act as energy harvesting relays to help the far users. The NNNF scheme can enable the selected near user to harvest more energy. Then we select a user within the ring D_A which has the shortest distance to the BS as the far NOMA user (denoted by A_{i*}). The advantage of NNNF scheme is that it can minimize the outage probability of both the near and far users.

Using the same definition of the outage probability as the near users of NOMA, we can characterize the outage probability of the near users of NNNF.

The following theorem provides the outage probability of the near users of NNNF for an arbitrary choice of α.

Theorem 4.3 *Conditioned on PPP, the outage probability of B_{i*} can be approximated in closed-form as follows:*

$$P_{B_{i*}} \approx b_1 \sum_{n=1}^{N} \sqrt{1 - \phi_n^2} \left(1 - e^{-(1+c_{n*}^\alpha)\varepsilon_{A_i}}\right) c_{n*} e^{-\pi \lambda_{\Phi_B} c_{n*}^2}, \tag{4.22}$$

if $\varepsilon_{A_i} \geq \varepsilon_{B_i}$, otherwise $P_{B_{i*}} = 1$, where $c_{n*} = \frac{R_{D_B}}{2}(\phi_n + 1)$, $b_1 = \frac{\xi_B \omega_N R_{D_B}}{2}$, and $\xi_B = \frac{2\pi \lambda_{\Phi_B}}{1 - e^{-\pi \lambda_{\Phi_B} R_{D_B}^2}}$.

After some manipulations, the following corollary can be obtained.

Corollary 4.3 *For the special case $\alpha = 2$, the outage probability of B_{i*} can be expressed in exact closed-form as follows:*

$$P_{B_{i*}}\Big|_{\alpha=2} = \frac{\xi_B \left(e^{-R_{D_B}^2 \left(\pi \lambda_{\Phi_B} + \varepsilon_{A_i}\right) - \varepsilon_{A_i}} - e^{-\varepsilon_{A_i}}\right)}{2\left(\pi \lambda_{\Phi_B} + \varepsilon_{A_i}\right)} - \frac{\xi_B \left(e^{-\pi \lambda_{\Phi_B} R_{D_B}^2} - 1\right)}{2\pi \lambda_{\Phi_B}}, \tag{4.23}$$

if $\varepsilon_{A_i} \geq \varepsilon_{B_i}$, otherwise $P_{B_{i*}}\big|_{\alpha=2} = 1$.

Using the same definition of the outage probability of the far users of NOMA, and similar to (4.16), we can characterize the outage probability of the far users of NNNF. The following theorem provides the outage probability of the far users of NNNF for an arbitrary choice of α.

Theorem 4.4 *Conditioned on PPP, assume $R_{DC} \gg R_{DB}$, the outage probability of A_{i*} can be approximated in closed-form as follows:*

$$
\begin{aligned}
P_{A_{i*}} \approx \varsigma^* & \sum_{n=1}^{N} \sqrt{1 - \phi_n{}^2} \left(1 + c_{n*}^{\alpha}\right) c_{n*} e^{-\pi \lambda_{\Phi_B} c_{n*}^2} \\
\times & \sum_{k=1}^{K} \sqrt{1 - \psi_k^2}(1 + s_k^{\alpha})^2 s_k e^{-\pi \lambda_{\Phi_A} \left(s_k^2 - R_{DC}^2\right)} \sum_{m=1}^{M} \sqrt{1 - \varphi_m^2} \\
\times & \; e^{-(1 + s_k^{\alpha}) t_m} \chi_{t_m} \left(\ln \frac{\chi_{t_m} \left(1 + s_k^{\alpha}\right)\left(1 + c_{n*}^{\alpha}\right)}{\eta \rho} + 2 c_0 \right) \\
+ & \, b_2 b_3 \sum_{k=1}^{K} \sqrt{1 - \psi_k^2}(1 + s_k^{\alpha}) s_k e^{-\pi \lambda_{\Phi_A} s_k^2} \\
\times & \sum_{n=1}^{N} \left(\sqrt{1 - \phi_n{}^2} \left(1 + c_{n*}^{\alpha}\right) c_{n*} e^{-\pi \lambda_{\Phi_B} c_{n*}^2} \right),
\end{aligned}
\tag{4.24}
$$

where $\varsigma^* = -\dfrac{\xi_B \xi_A \omega_N \omega_K \omega_M \varepsilon_{A_i} R_{D_B} (R_{D_A} - R_{D_C})}{8 \eta \rho}$, $b_2 = \dfrac{\xi_A e^{\pi \lambda_{\Phi_A} R_{D_C}^2} \omega_K \varepsilon_{A_i}}{R_{D_A} + R_{D_C}}$, *and* $b_3 = \dfrac{\xi_B \omega_N R_{D_B} \varepsilon_{A_i}}{2}$.

Corollary 4.4 *For the special case $\alpha = 2$, the outage probability of A_{i*} can be simplified in closed-form as (4.25) on the top of this page.*

$$
\begin{aligned}
P_{A_{i*}} \big|_{\alpha=2} \approx \varsigma^* & \sum_{n=1}^{N} \sqrt{1 - \phi_n{}^2} \left(1 + c_{n*}^2\right) c_{n*} e^{-\pi \lambda_{\Phi_B} c_{n*}^2} \\
\times & \sum_{k=1}^{K} \sqrt{1 - \psi_k^2}\left(1 + s_k^2\right)^2 s_k e^{-\pi \lambda_{\Phi_A} \left(s_k^2 - R_{DC}^2\right)} \\
\times & \sum_{m=1}^{M} \sqrt{1 - \varphi_m^2} \left(e^{-(1 + s_k^2) t_m} \chi_{t_m} \left(\ln \frac{\chi_{t_m} \left(1 + s_k^2\right)\left(1 + c_{n*}^2\right)}{\eta \rho} + 2 c_0 \right) \right) \\
+ & \, \frac{\xi_A e^{\pi \lambda_{\Phi_A} R_{DC}^2}}{2} \left(\frac{e^{-\varepsilon_{A_i}}}{\pi \lambda_{\Phi_A} + \varepsilon_{A_i}} \left(e^{-R_{DA}^2 \left(\pi \lambda_{\Phi_A} + \varepsilon_{A_i}\right)} - e^{-R_{DC}^2 \left(\pi \lambda_{\Phi_A} + \varepsilon_{A_i}\right)} \right) \right.
\end{aligned}
$$

$$-\frac{\left(e^{-\pi\lambda_{\Phi_A}R_{D_A}^2} - e^{-\pi\lambda_{\Phi_A}R_{D_C}^2}\right)}{\pi\lambda_{\Phi_A}}\Bigg)$$

$$\times\frac{\xi_B}{2}\left(\frac{e^{-R_{D_B}^2\left(\pi\lambda_{\Phi_B}+\varepsilon_{A_i}\right)-\varepsilon_{A_i}} - e^{-\varepsilon_{A_i}}}{\pi\lambda_{\Phi_B}+\varepsilon_{A_i}} - \frac{e^{-\pi\lambda_{\Phi_B}R_{D_B}^2}-1}{\pi\lambda_{\Phi_B}}\right).$$

$$(4.25)$$

Similarly, we provide diversity analysis of both the near and far users of NNNF.

Near Users For the near users, based on the analytical results, we carry out the high SNR approximation as follows. When $\varepsilon \to 0$, a high SNR approximation of (4.22) with $1 - e^{-x} \approx x$ is given by

$$P_{B_{j*}} \approx b_1\varepsilon_{A_i}\sum_{n=1}^{N}\left(\sqrt{1-\phi_n^2}\left(1+c_{n*}^\alpha\right)c_{n*}e^{-\pi\lambda_{\Phi_B}c_{n*}^2}\right).\qquad(4.26)$$

Substituting (4.26) into (4.20), we obtain that the diversity gain for the near users of NNNF is one, which indicates that using NNNF will not affect the diversity gain.

Far User For the far users, substituting (4.24) into (4.20), we obtain that the diversity gain is still two. It indicates that NNNF will not affect the diversity gain.

4.2.2.3   NNFF Selection Scheme

In this scheme, we first select a user within disc D_B which has the shortest distance to the BS as a near NOMA user. Then we select a user within ring D_A which has the farthest distance to the BS as a far NOMA user (denoted by $A_{i'}$). The use of this selection scheme is inspired by an interesting observation described in Ding et al. (2014b) that NOMA can offer a larger performance gain over conventional MA when user channel conditions are more different.

Since the same criterion for the near users is used, the outage probabilities of near nears for an arbitrary α and the special case $\alpha = 2$ are the same as those expressed in (4.22) and (4.23), respectively.

Using the same definition of the outage probability of the far users, and similar to (4.16), we can characterize the outage probability of the far users of NNFF. The following theorem provides the outage probability of the far user of NNFF for an arbitrary choice of α.

Theorem 4.5 *Conditioned on PPP, assume $R_{D_C} \gg R_{D_B}$, the outage probability of $A_{i'}$ can be approximated as follows:*

$$P_{A_{i'}} \approx \varsigma^* \sum_{n=1}^{N} \sqrt{1 - \phi_n^2} \left(1 + c_{n*}^{\alpha}\right) c_{n*} e^{-\pi \lambda_{\Phi_B} c_{n*}^2}$$

$$\times \sum_{k=1}^{K} \sqrt{1 - \psi_k^2} (1 + s_k^{\alpha})^2 s_k e^{-\pi \lambda_{\Phi_A} \left(R_{D_A}^2 - s_k^2\right)} \sum_{m=1}^{M} \sqrt{1 - \varphi_m^2}$$

$$\times e^{-(1+s_k^{\alpha}) t_m} \chi_{t_m} \left(\ln \frac{\chi_{t_m} \left(1 + s_k^{\alpha}\right) \left(1 + c_{n*}^{\alpha}\right)}{\eta \rho} + 2c_0 \right)$$

$$+ b_3 b_4 \sum_{k=1}^{K} \sqrt{1 - \psi_k^2} (1 + s_k^{\alpha}) s_k e^{\pi \lambda_{\Phi_A} s_k^2}$$

$$\times \sum_{n=1}^{N} \left(\sqrt{1 - \phi_n^2} \left(1 + c_{n*}^{\alpha}\right) c_{n*} e^{-\pi \lambda_{\Phi_B} c_{n*}^2} \right), \tag{4.27}$$

where $b_4 = \dfrac{\xi_A e^{-\pi \lambda_{\Phi_A} R_{D_A}^2} \omega_K \varepsilon_{A_i}}{R_{D_A} + R_{D_C}}$.

Corollary 4.5 *For the special case $\alpha = 2$, after some manipulations, the outage probability of $A_{i'}$ can be simplified in closed-form as (4.28) on the top of the next page.*

$$P_{A_{i'}}\big|_{\alpha=2} \approx \varsigma^* \sum_{n=1}^{N} \sqrt{1 - \phi_n^2} \left(1 + c_{n*}^2\right) c_{n*} e^{-\pi \lambda_{\Phi_B} c_{n*}^2}$$

$$\times \sum_{k=1}^{K} \sqrt{1 - \psi_k^2} \left(1 + s_k^2\right)^2 s_k e^{-\pi \lambda_{\Phi_A} \left(R_{D_A}^2 - s_k^2\right)}$$

$$\times \sum_{m=1}^{M} \sqrt{1 - \varphi_m^2} \left(e^{-(1+s_k^2) t_m} \chi_{t_m} \left(\ln \frac{\chi_{t_m} \left(1 + s_k^2\right) \left(1 + c_{n*}^2\right)}{\eta \rho} + 2c_0 \right) \right)$$

$$+ \frac{\xi_A e^{-\pi \lambda_{\Phi_A} R_{D_A}^2}}{2} \left(\frac{e^{\pi \lambda_{\Phi_A} R_{D_A}^2} - e^{\pi \lambda_{\Phi_A} R_{D_C}^2}}{\pi \lambda_{\Phi_A}} - \frac{e^{-\varepsilon_{A_i}}}{\pi \lambda_{\Phi_A} - \varepsilon_{A_i}} \right.$$

$$\left. \times \left(e^{R_{D_A}^2 (\pi \lambda_{\Phi_A} - \varepsilon_{A_i})} - e^{R_{D_C}^2 (\pi \lambda_{\Phi_A} - \varepsilon_{A_i})} \right) \right)$$

$$\times \frac{\xi_B}{2} \left(\frac{\left(e^{-R_{D_B}^2 (\pi \lambda_{\Phi_B} + \varepsilon_{A_i}) - \varepsilon_{A_i}} - e^{-\varepsilon_{A_i}} \right)}{(\pi \lambda_{\Phi_B} + \varepsilon_{A_i})} - \frac{\left(e^{-\pi \lambda_{\Phi_B} R_{D_B}^2} - 1 \right)}{\pi \lambda_{\Phi_B}} \right). \tag{4.28}$$

Similarly, we provide diversity analysis of both the near and far uses of NNFF.

Near Users Since the same criterion for selecting a near user is used, the diversity gain is one which is the same as NNNF.

Far Users Substituting (4.27) into (4.20), we find that the diversity gain is still two. Therefore, we conclude that using opportunistic user selection schemes (NNNF and NNFF) based on distances will not affect the diversity gain.

4.2.3 Numerical Results

In this section, numerical results are presented to facilitate the performance evaluations (including the outage probability of the near and the far users and the delay sensitive throughput) of the proposed cooperative SWIPT NOMA protocol. In the considered network, we assume that the energy conversion efficiency of SWIPT is $\eta = 0.7$ and the power allocation coefficients of NOMA is $|p_{i1}|^2 = 0.8$, $|p_{i1}|^2 = 0.2$. In the following figures, we use red, blue, and black color lines to represent the RNRF, NNNF, and NNFF user selection schemes, respectively.

4.2.3.1 Outage Probability of the Near Users

In this subsection, the outage probability achieved by the near users with different choice of density and path loss coefficients for the three user selection schemes is demonstrated. Note that the same user selection criteria is applied for the near users of NNNF and NNFF, and we use NNN(F)F to represent these two selection schemes in Figs. 4.3 and 4.4.

Fig. 4.3 Outage probability of the near users versus SNR with different α, where $R_{D_B} = 2\,m$, and $\lambda_{\Phi_B} = 1$

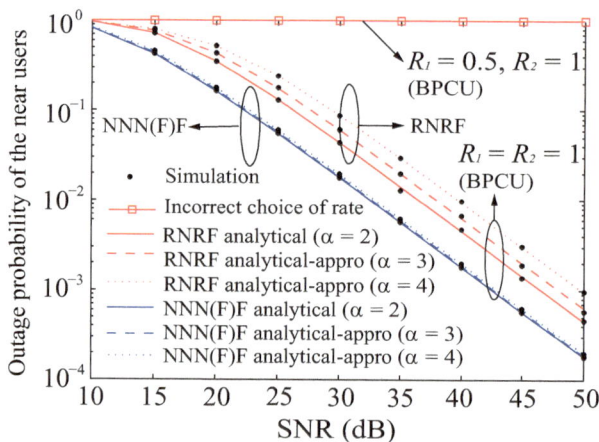

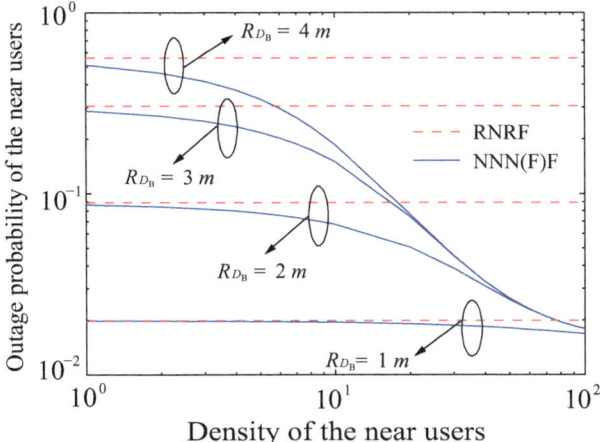

Fig. 4.4 Outage probability of the near users versus density with different R_{D_B}, where $\lambda_{\Phi_B} = 1$, and $SNR = 30$ dB

Figure 4.3 plots the outage probability of the near users versus SNR with different path loss coefficients for both RNRF and NNN(F)F. The solid red and blue curves are for the special case $\alpha = 2$ of RNRF and NNN(F)F, corresponding to the analytical results derived in (4.15) and (4.23), respectively. The dashed red and blue curves are for an arbitrary choice of α, corresponding to the analytical results derived in (4.14) and (4.22), respectively. Monte Carlo simulation are marked as "•" to verify our derivation. The figure shows precise agreement between the simulation and analytical curves. One can observe that by performing NNNF and NNFF (which we refer to as NNN(F)F in the figure), lower outage probability is achieved than RNRF since shorter distance makes less path loss and leads to better performance. The figure also demonstrates that as α increases, outage will occur more because of higher path loss. For NNNF and NNFF, the performance is very close for different values of α. This is because we use the bounded path loss model (i.e., $1 + d_i^\alpha > 1$) to ensure that the path loss is always larger than one. When selecting the nearest near user, d_i will approach to zero and the path loss will approach to one, which makes the performance difference of the three selection schemes insignificant. It is worth noting that all curves have the same slopes, which indicates that the diversity gains of the schemes are the same. This phenomenon validates the insights we obtained from the analytical results derived in (4.20). Figure 4.3 also shows that if the choices of rates for users are incorrect (i.e., $R_1 = 0.5$ and $R_2 = 1$ in this figure), the outage probability of the near users will be always one, which verifies the analytical results in (4.14) and (4.22).

Figure 4.4 plots the outage probability of the near users versus the density with different R_{D_B}. RNRF is also shown in the figure as a benchmark for comparison. Several observations are drawn as follows: (1) The outage probabilities of RNRF and NNN(F)F decrease by decreasing R_{D_B} because path loss is reduced; (2) The outage probability of NNN(F)F decreases as the density of the near users increases. This is due to the multiuser diversity gain, since there is an increasing number of the near users; (3) The outage probability of RNRF is a constant, i.e., independent of the density of near users, and is the outage ceiling of the NNN(F)F. This is due to the fact that no opportunistic user selection is carried out for RNRF; and (4) An outage floor exits even if the density of the near users goes to infinity. This is due to the used bounded path loss model. When the number of the near users exceeds a threshold, the selected near user will be very close to the source, which makes the path loss approach to one.

4.2.3.2 Outage Probability of the Far Users

In this subsection, we demonstrate the outage probability of the far users with different choices of the density, path loss coefficients, and user zone of the three user selection schemes.

Figure 4.5 plots the outage probability of the far users versus SNR with different path loss coefficients of RNRF, NNNF, and NNFF. The dashed red, blue, and black curves circled together and pointed by $\alpha = 2$ are the analytical approximation for the special case of RNRF, NNNF, and NNFF, which are obtained from (4.18), (4.25), and (4.28), respectively. The dashed red, blue, and black curves circled together and pointed by $\alpha = 3$ are the analytical approximation for an arbitrary choice of α of RNRF, NNNF, and NNFF, which are obtained from (4.17), (4.24), and (4.27), respectively. We use the solid marked lines to represent the Monte Carlo simulation results for each case. As can be observed from the figure, the simulation and the analytical approximation are very close, particularly in the high SNR region. Several

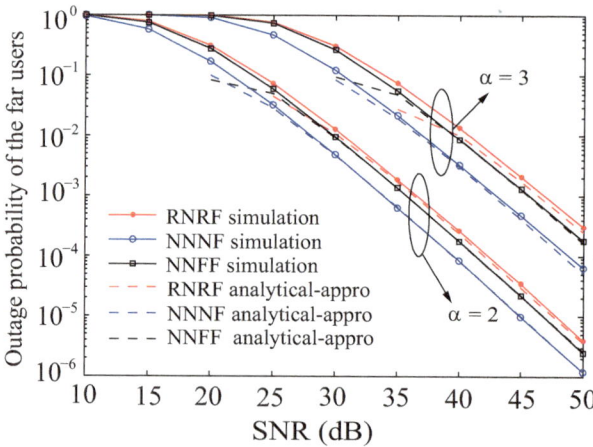

Fig. 4.5 Outage probability of the far users with different α, $R_1 = 0.3$ BPCU, $R_{D_A} = 10$ m, $R_{D_B} = 2$ m, $R_{D_C} = 8$ m, $\lambda_{\varPhi_A} = 1$, and $\lambda_{\varPhi_B} = 1$

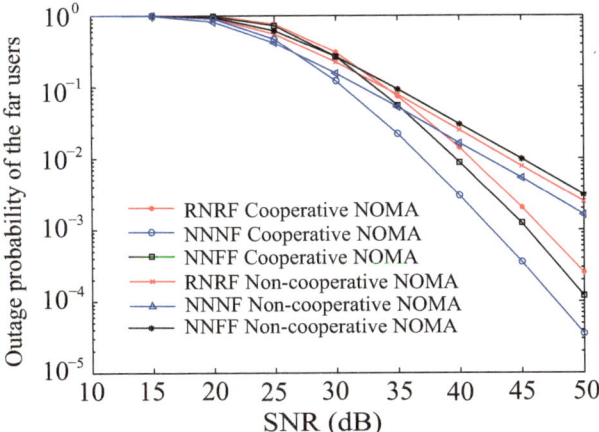

Fig. 4.6 Comparison of outage probability with noncooperative NOMA, $\alpha = 3$, $R_1 = 0.3$ BPCU, $R_{D_A} = 10\,\text{m}$, $R_{D_B} = 2\,\text{m}$, $R_{D_C} = 8\,\text{m}$, $\lambda_{\Phi_A} = 1$, and $\lambda_{\Phi_B} = 1$

observations are drawn as follows: (1) NNNF achieves the lowest outage probability among the three selection schemes since both the near and far users have the smallest path loss; (2) NNFF achieves lower outage than RNRF, which indicates that the distance of the near users has more impact than that of the far users; (3) it is clear that all of the curves in Fig. 4.5 have the same slopes, which indicates that the diversity gain of the far users for the three schemes is the same. In the diversity analysis part, we derive the diversity gain of the three selection schemes is two. The simulation validates the analytical results and indicates that the achievable diversity gain is the same for different user selection schemes.

Figure 4.6 plots the outage probability of the far users versus SNR for both cooperative NOMA and noncooperative NOMA.[1] Several observations are drawn as follows: (1) by using an energy constrained relay to perform cooperative NOMA transmission, the outage probability of the far users has a larger slope than that of noncooperative NOMA, for all user selection schemes. This is due to the fact that cooperative NOMA can achieve a larger diversity gain and guarantees more reliable reception for the far users in high SINR region; (2) NNNF achieves the lowest outage probability among these three selection schemes both for cooperative NOMA and noncooperative NOMA because of its smallest path loss; (3) it is worth noting that NNFF has higher outage probability than RNRF in noncooperative NOMA; however, it achieves lower outage probability than RNRF in cooperative

---

[1]It is common to use outage probability as a criterion to compare the performance of cooperative transmission and noncooperative transmission schemes (Laneman et al. 2004). In the context of cooperative NOMA, the use of outage probability is particularly useful since cooperative NOMA is to improve the reception reliability of the far users.

NOMA. This phenomenon indicates that it is very helpful and necessary to apply cooperative NOMA in NNFF due to the largest performance gain over noncooperative NOMA.

4.3 Summary

This chapter discusses the sustainability of NOMA by illustrating the cooperative NOMA and the performance analysis of wireless powered NOMA networks for lifetime extension.

References

Ali, M. S., Hossain, E., & Kim, D. I. (2017). Coordinated multi-point (CoMP) transmission in downlink multi-cell NOMA systems: Models and spectral efficiency performance. arXiv preprint. arXiv:1703.09255.

Choi, J. (2014). Non-orthogonal multiple access in downlink coordinated two-point systems. *IEEE Communications Letters, 18*, 313–316.

Cover, T. M., & Thomas, J. A. (2006). *Elements of information theory* (2nd ed.). Hoboken: Wiley.

Ding, Z., Dai, H., & Poor, H. V. (2016). Relay selection for cooperative NOMA. *IEEE Communications Letters, 5*, 416–419.

Ding, Z., Krikidis, I., Sharif, B., & Poor, H. V. (2014a). Wireless information and power transfer in cooperative networks with spatially random relays. *IEEE Transactions on Wireless Communications, 13*(8), 4440–4453, Aug. 2014.

Ding, Z., Peng, M., & Poor, H. V. (2015). Cooperative non-orthogonal multiple access in 5G systems. *IEEE Communications Letters, 19*, 1462–1465.

Ding, Z., & Poor, H. (2013). Cooperative energy harvesting networks with spatially random users. *IEEE Signal Processing Letters, 20*, 1211–1214.

Ding, Z., Yang, Z., Fan, P., & Poor, H. V. (2014b). On the performance of non-orthogonal multiple access in 5G systems with randomly deployed users. *IEEE Signal Processing Letters, 21*, 1501–1505.

Duan, W., Wen, M., Yan, Y., Xiong, Z., & Lee, M. H. (2016). Use of non-orthogonal multiple access in dual-hop relaying. arXiv preprint arXiv:1604.01151.

Kim, J. B., & Lee, I. H. (2015). Non-orthogonal multiple access in coordinated direct and relay transmission. *IEEE Communications Letters, 19*, 2037–2040.

Krikidis, I., Sasaki, S., Timotheou, S., & Ding, Z. (2014). A low complexity antenna switching for joint wireless information and energy transfer in MIMO relay channels. *IEEE Transactions on Communications, 62*, 1577–1587.

Laneman, J. N., Tse, D. N., & Wornell, G. W. (2004). Cooperative diversity in wireless networks: Efficient protocols and outage behavior. *IEEE Transactions on Information Theory, 50*, 3062–3080.

Liu, Y., Ding, Z., Elkashlan, M., & Poor, H. V. (2016). Cooperative non-orthogonal multiple access with simultaneous wireless information and power transfer. *IEEE Journal on Selected Areas in Communications, 34*.

Martin-Vega, F., Liu, Y., Gomez, G., Aguayo-Torres, M. C., & Elkashlan, M. (2017). Modeling and analysis of NOMA enabled CRAN with cluster point process. Piscataway: IEEE.

Men, J., & Ge, J. (2015a). Non-orthogonal multiple access for multiple-antenna relaying networks. *IEEE Communications Letters, 19*, 1686–1689.

Men, J., & Ge, J. (2015b). Performance analysis of non-orthogonal multiple access in downlink cooperative network. *IET Communications, 9,* 2267–2273.

Nasir, A. A., Zhou, X., Durrani, S., & Kennedy, R. A. (2013). Relaying protocols for wireless energy harvesting and information processing. *IEEE Transactions on Wireless Communications, 12,* 3622–3636.

Qin, Z., Liu, Y., Gao, Y., Elkashlan, M., & Nallanathan, A. (2017). Wireless powered cognitive radio networks with compressive sensing and matrix completion. *IEEE Transactions on Communications, 65,* 1464–1476.

Shin, W., Vaezi, M., Lee, B., Love, D. J., Lee, J., & Poor, H. V. (2017). Coordinated beamforming for multi-cell MIMO-NOMA. *IEEE Communications Letters, 21,* 84–87.

Tian, Y., Nix, A., & Beach, M. (2016). On the performance of opportunistic NOMA in downlink CoMP networks. *IEEE Communications Letters, 20,* 998–1001.

Timotheou, S., & Krikidis, I. (2015). Fairness for non-orthogonal multiple access in 5G systems. *IEEE Signal Processing Letters,22,* 1647–1651.

Varshney, L. (2008). Transporting information and energy simultaneously. In *Proceedings of IEEE International Symposium on Information Theory (ISIT)*, Toronto, ON (pp. 1612–1616).

Vien, Q. T., Ogbonna, N., Nguyen, H. X., Trestian, R., & Shah, P. (2015). Non-orthogonal multiple access for wireless downlink in cloud radio access networks. In *Proceedings of European Wireless Conference* (pp. 1–6).

Yang, K., Yang, N., Ye, N., Jia, M., Gao, Z., & Fan, R. (2019). Non-orthogonal multiple access: Achieving sustainable future radio access. *IEEE Communications Magazine, 57,* 116–121.

Zhang, R., & Ho, C. K. (2013). MIMO broadcasting for simultaneous wireless information and power transfer. *IEEE Transactions on Communications, 12,* 1989–2001.

Chapter 5
Security in NOMA

In this chapter, the security of NOMA will be demonstrating by introducing average performance of physical layer security (PLS).

5.1 Secure NOMA in Random Wireless Networks

Given the broadcast nature of wireless transmissions, the concept of physical (PHY) layer security (PLS), which was proposed by Wyner as early as 1975 from an information-theoretical perspective (Wyner 1975), has sparked of widespread recent interest. To elaborate, PLS has been considered from different perspectives. Specifically, robust beamforming transmission was conceived in conjunction with applying artificial noise (AN) for mitigating the impact of imperfect channel state information (CSI) in MIMO wiretap channels was proposed by Mukherjee and Swindlehurst (2011). The security issue in CRNs has been investigated by Qin et al. (2018). Furthermore, the physical layer security of D2D communication in large-scale CRNs was investigated by Liu et al. (2016) by invoking wireless power transfer and the design of decoding order, transmission rates, and power allocation in the secure NOMA has been investigated by He et al. (2017). The relay selection has been investigated to enhance the security in NOMA networks by Feng et al. (2019).

Recently, various PHY layer techniques such as cooperative jamming (Tekin and Yener 2008) and AN (Goel and Negi 2008) aided solutions were proposed for improving the PLS, even if the eavesdroppers have better channel conditions than the legitimate receivers. A popular technique is to generate AN at the transmitter for degrading the eavesdroppers' reception, which was proposed by Goel and Negi (2008). In contrast to the traditional view, which regards noise and interference as a detrimental effect, generating AN at the transmitter is capable of improving the security, because it degrades the channel conditions of eavesdroppers without

© The Author(s), under exclusive license to Springer Nature Switzerland AG 2020   67
Y. Liu et al., *Non-Orthogonal Multiple Access for Massive Connectivity*,
SpringerBriefs in Computer Science, https://doi.org/10.1007/978-3-030-30975-6_5

Fig. 5.1 Network model for
secure NOMA transmission
in single-antenna scenario

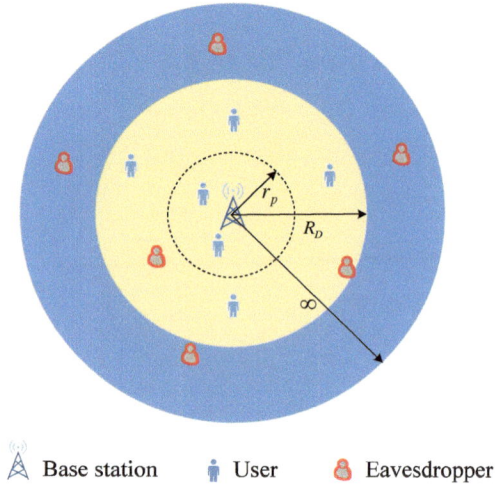

affecting those of the legitimate receivers. An AN-based multi-antenna-aided secure
transmission scheme affected by colluding eavesdroppers was considered by Zhou
and McKay (2010) for the scenarios associated both with perfect and imperfect
CSI at both the transmitter and receiver. As a further development, the secrecy
enhancement achieved in wireless Ad Hoc networks was investigated by Zhang
et al. (2013), with the aid of both beamforming and sectoring techniques. In this
chapter, we specifically consider the scenario of large-scale networks, where a BS
supports randomly roaming NOMA users.

5.1.1 System Model

As shown in Fig. 5.1, we focus our attention on a secure downlink communication
scenario. In the scenario considered, a BS communicates with M legitimate users
(LUs) in the presence of eavesdroppers (Es). We assume that the M users are divided
into $M/2$ orthogonal pairs. For each pair, the NOMA transmission protocol is
invoked. It is assumed that BS is located at the center of a disc, denoted by $\mathscr{D}$, which
has a coverage radius of R_D, which is defined as the user zone for NOMA (Ding
et al. 2014). The M randomly roaming LUs are uniformly distributed within the
disc. A random number of Es is distributed in an infinite two-dimensional plane.
The spatial distribution of all Es is modeled using a homogeneous PPP, which is
denoted by Φ_e associated with the density λ_e. It is assumed that the Es can be
detected, provided that they are close enough to BS. Therefore, an E-exclusion area
having a radius of r_p is introduced. Additionally, all channels are assumed to impose
quasi-static Rayleigh fading, where the channel coefficients are constant for each
transmission block, but vary independently between different blocks.

Without loss of generality, it is assumed that all the channels between the BS and LUs obey $|h_1|^2 \leq \cdots |h_m|^2 \leq \cdots |h_n|^2 \leq \cdots |h_M|^2$. Both the small-scale fading and the path loss are incorporated into the ordered channel gain. Again, we assume that the m-th user and the n-th user ($m < n$) are paired for transmission in the same resource slot. With loss of generality, we focus our attention on a single selected pair of users in the rest of the chapter. In the NOMA transmission protocol, more power should be allocated to the user suffering from worse channel condition (Ding et al. 2017; Saito et al. 2013). Therefore, the power allocation coefficients satisfy the conditions that $a_m \geq a_n$ and $a_m + a_n = 1$. SIC is invoked for detecting the stronger user first. Based on the aforementioned assumptions, the instantaneous SINR of m-th user and the signal-to-noise ratio (SNR) of n-th user can be written as:

$$\gamma_{B_m} = \frac{a_m |h_m|^2}{a_n |h_m|^2 + \frac{1}{\rho_b}},\tag{5.1}$$

and

$$\gamma_{B_n} = \rho_b a_n |h_n|^2,\tag{5.2}$$

respectively. We introduce the convenient concept of transmit SNR $\rho_b = \frac{P_T}{\sigma_b^2}$, where P_T is the transmit power at the BS and σ_b^2 is the variance of the additive white Gaussian noise (AWGN) at the LUs, noting that this is not a physically measurable quantity owing to their geographic separation. In order to ensure that the m-th user can successfully decode the message of the n-th user, the condition of $a_m \geq \left(2^{R_m} - 1\right) a_n$ should be satisfied. Additionally, a bounded path loss model is used for guaranteeing that there is a practical path-loss, which is higher than one even for small distances.

We consider the worst-case scenario of large-scale networks, in which the Es are assumed to have strong detection capabilities. Specifically, by applying multiuser detection techniques, the multiuser data stream received from BS can be distinguished by the Es. In the scenario considered, all the downlink CSIs are assumed to be known at BS. Under this assumption, the most detrimental E is not necessarily the nearest one, but the one having the best channel to BS. Therefore, the instantaneous SNR of detecting the information of the m-th user and the n-th user at the most detrimental E can be expressed as follows:

$$\gamma_{E_\kappa} = \rho_e a_\kappa \max_{e \in \Phi_e, d_e \geq r_p} \left\{ |g_e|^2 L(d_e) \right\}.\tag{5.3}$$

It is assumed that $\kappa \in \{m, n\}$, $\rho_e = \frac{P_A}{\sigma_e^2}$ is the transmit SNR with σ_e^2 being the variance of the AWGN at Es. Additionally, g_e is defined as the small-scale fading coefficient associated with $g_e \sim \mathcal{CN}(0, 1)$, $L(d_e) = \frac{1}{d_e^\alpha}$ is the path loss, and d_e is the distance from Es to BS.

5.1.2   New Channel Statistics

In this subsection, we derive several new channel statistics for LUs and Es, which will be used for deriving the secrecy outage probability in the next subsection.

Lemma 5.1 *Assuming M randomly located NOMA users in the disc of Fig. 5.1, the cumulative distribution function (CDF) $F_{\gamma_{B_n}}$ of the n-th LU is given by*

$$
F_{\gamma_{B_n}}(x) = \varphi_n \sum_{p=0}^{M-n} \binom{M-n}{p} \frac{(-1)^p}{n+p} \sum_{\tilde{S}_n^p} \binom{n+p}{q_0 + \cdots + q_K} \left(\prod_{K=0}^{K} b_k^{q_k} \right)
$$

$$
\times e^{-\sum_{k=0}^{K} q_k c_k \frac{x}{\rho_b a_n}},
\tag{5.4}
$$

where K is a complexity-vs-accuracy tradeoff parameter, $b_k = -\omega_K \sqrt{1 - \phi_k^2}$ $(\phi_k + 1)$, $b_0 = -\sum_{k=1}^{K} b_k$, $c_k = 1 + \left[\frac{R_D}{2}(\phi_k + 1) \right]^\alpha$, $\omega_K = \frac{\pi}{K}$, and $\phi_k = \cos\left(\frac{2k-1}{2K} \pi \right)$, $\tilde{S}_n^p = \left\{ (q_0, q_1, \ldots, q_K) \mid \sum_{i=0}^{K} q_i = n + p \right\}$, $\binom{n+p}{q_0 + \cdots + q_K} = \frac{(n+p)!}{q_0! \cdots q_K!}$ and $\varphi_n = \frac{M!}{(M-n)!(n-1)!}$.

Lemma 5.2 *Assuming M randomly positioned NOMA users in the disc of Fig. 5.1, the CDF $F_{\gamma_{B_m}}$ of the m-th LU is given in (5.5)*

$$
F_{\gamma_{B_m}}(x) = U\left(x - \frac{a_m}{a_n} \right) + U\left(\frac{a_m}{a_n} - x \right) \varphi_m
$$

$$
\times \sum_{p=0}^{M-m} \binom{M-m}{p} \frac{(-1)^p}{m+p} \sum_{\tilde{S}_m^p} \binom{m+p}{q_0 + \cdots + q_K} \left(\prod_{k=0}^{K} b_k^{q_k} \right)
$$

$$
\times e^{-\sum_{k=0}^{K} q_k c_k \frac{x}{(a_m - a_n x)\rho_b}}.
\tag{5.5}
$$

where $U(x)$ is the unit step function formulated as $U(x) = \begin{cases} 1, x > 0 \\ 0, x \leq 0 \end{cases}$, and $\tilde{S}_m^p = \left\{ (q_0, q_1, \ldots, q_K) \mid \sum_{i=0}^{K} q_i = m + p \right\}$.

Lemma 5.3 *Assuming that the eavesdroppers obey the PPP distribution and the E-exclusion zone has a radius of r_p, the probability density function (PDF) $f_{\gamma_{E_K}}$ of the most detrimental E (where $\kappa \in \{m, n\}$) is given by*

$$f_{\gamma_{E_\kappa}}(x) = \mu_{\kappa 1}e^{-\frac{\mu_{\kappa 1}\Gamma(\delta,\mu_{\kappa 2}x)}{x^\delta}}\left(\frac{\mu_{\kappa 2}^\delta e^{-\mu_{\kappa 2}x}}{x} + \frac{\delta\Gamma(\delta,\mu_{\kappa 2}x)}{x^{\delta+1}}\right), \tag{5.6}$$

where $\mu_{\kappa 1} = \delta\pi\lambda_e(\rho_e a_\kappa)^\delta$, $\mu_{\kappa 2} = \frac{r_p^\alpha}{\rho_e a_\kappa}$, $\delta = \frac{2}{\alpha}$, and $\Gamma(\cdot,\cdot)$ is the upper incomplete Gamma function.

5.1.3 Secrecy Outage Probability

In this chapter, the SOP is used as our secrecy performance metric. Additionally, the secrecy rate of the m-th and of the n-th user can be expressed as

$$I_m = \left[\log_2(1 + \gamma_{B_m}) - \log_2(1 + \gamma_{E_m})\right]^+, \tag{5.7}$$

and

$$I_n = \left[\log_2(1 + \gamma_{B_n}) - \log_2(1 + \gamma_{E_n})\right]^+, \tag{5.8}$$

respectively, where we have $[x]^+ = \max\{x, 0\}$. Given the expected secrecy rate R_m and R_n for the m-th and n-th users, a secrecy outage event is declared, when the instantaneous secrecy rate drops below R_m and R_n, respectively. Based on (5.7), the SOP for the m-th user is given by

$$P_m(R_m) = \int_0^\infty f_{\gamma_{E_m}}(x) F_{\gamma_{B_m}}\left(2^{R_m}(1+x) - 1\right)dx. \tag{5.9}$$

Based on the assumption of $a_m \geq \left(2^{R_m} - 1\right)a_n$, we consider the SOP under the condition that the connection between BS and LUs can be established. Upon using the results of Lemmas 5.2 and 5.3, as well as substituting (5.5) and (5.6) into (5.9), after some further mathematical manipulations, we can express the SOP of the m-th user according to the following theorem:

Theorem 5.1 *Assuming that the LUs position obeys the PPP for the ordered channels of the LUs, the SOP of the m-th user is given by (5.10)*

$$P_m(R_m) = 1 - e^{-\frac{\mu_{m1}\Gamma(\delta,\tau_m\mu_{m2})}{\tau_m^\delta}} + \varphi_m\sum_{p=0}^{M-m}\binom{M-m}{p}\frac{(-1)^p}{m+p}$$

$$\times \sum_{\tilde{S}_m^p}\binom{m+p}{q_0+\cdots+q_K}\left(\prod_{k=0}^K b_k^{q_k}\right)$$

$$\times \int_0^{\tau_m} \mu_{m1} \left(\frac{\mu_{m2}^{\delta} e^{-\mu_{m2}x}}{x} + \frac{\delta \Gamma\left(\delta, \mu_{m2}x\right)}{x^{\delta+1}} \right)$$

$$\times e^{-\frac{\mu_{m1}\Gamma(\delta,\mu_{m2}x)}{x^{\delta}} - \sum_{k=0}^{K} q_k c_k \frac{2^{Rm}(1+x)-1}{(a_m - a_n(2^{Rm}(1+x)-1))\rho_b}} dx. \tag{5.10}$$

where we have $\tau_m = \frac{1}{2^{Rm}(1-a_m)} - 1.$

Similarly, for the n-th user, based on (5.8), the SOP is given by

$$P_n\left(R_n\right) = \int_0^{\infty} f_{\gamma_{En}}(x) F_{\gamma_{Bn}}\left(2^{R_n}(1+x) - 1\right) dx. \tag{5.11}$$

Upon using the results of Lemmas 5.1 and 5.3, and substituting (5.4) and (5.6) into (5.11), after some further mathematical manipulations, we can express the SOP of the n-th user with the aid of the following theorem:

Theorem 5.2 *Assuming that the LUs position obeys the PPP for the ordered channels of the LUs, the SOP of the n-th user is given by*

$$P_n\left(R_n\right) = \varphi_n \sum_{p=0}^{M-n} \binom{M-n}{p} \frac{(-1)^p}{n+p} \sum_{\tilde{S}_n^p} \binom{n+p}{q_0 + \cdots + q_K} \left(\prod_{K=0}^{K} b_k^{q_k} \right)$$

$$\times \int_0^{\infty} \mu_{n1} \left(\frac{\mu_{n2}^{\delta} e^{-\mu_{n2}x}}{x} + \frac{\delta \Gamma\left(\delta, \mu_{n2}x\right)}{x^{\delta+1}} \right)$$

$$\times e^{-\frac{\mu_{n1}\Gamma(\delta,\mu_{n2}x)}{x^{\delta}} - \sum_{k=0}^{K} q_k c_k \frac{2^{Rn}(1+x)-1}{\rho_b a_n}} dx. \tag{5.12}$$

In this work, we consider that secrecy outage occurs in the m-th user and the n-th user are independent with each other. In other words, the SOP of the m-th user has an effect on the SOP of the n-th user and vice versa. As a consequence, we define the SOP for the selected user pair as that of either the m-th user or the n-th user outage. Hence, based on (5.10) and (5.12), the SOP of the selected user pair is given by

$$P_{mn} = 1 - (1 - P_m)(1 - P_n). \tag{5.13}$$

5.1.4 Secrecy Diversity Order Analysis

In order to derive the secrecy diversity order to gain further insights into the system's operation in the high-SNR regime, the following new analytical framework is introduced. Again, as the worst-case scenario, we assume that Es have a powerful detection capability. The asymptotic behavior is analyzed. Usually when the SNR of the channels between the BS and LUs is sufficiently high, i.e., when the BS's transmit SNR obeys $\rho_b \to \infty$, the SNR of the channels between BS and Es is set to arbitrary values. It is noted that for the E-transmit SNR of $\rho_e \to \infty$, the probability of successful eavesdropping will tend to unity. The secrecy diversity order can be defined as follows:

$$d_s = -\lim_{\rho_b \to \infty} \frac{\log P^\infty}{\log \rho_b}, \tag{5.14}$$

where P^∞ is the asymptotic SOP. We commence our diversity order analysis by characterizing the CDF of the LUs $F_{\gamma_{B_m}}^\infty$ and $F_{\gamma_{B_n}}^\infty$ in the high-SNR regime. When $y \to 0$, based on the approximation of $1 - e^{-y} \approx y$, we obtain the asymptotic unordered CDF of $\left|\tilde{h}_n\right|^2$ as follows:

$$F_{\left|\tilde{h}_n\right|^2}^\infty (y) \approx \frac{2y}{R_D^2} \int_0^{R_D} \left(1 + r^\alpha\right) r\, dr = y\ell, \tag{5.15}$$

where $\ell = 1 + \frac{2R_D^\alpha}{\alpha+2}$.

Based on (5.15), the asymptotic unordered CDF of $\left|\tilde{h}_n\right|^2$ is given by

$$F_{\left|\tilde{h}_n\right|^2}^\infty (y) = \varphi_n \sum_{p=0}^{M-n} \binom{M-n}{p} \frac{(-1)^p}{n+p} (y\ell)^{n+p} \approx \frac{\varphi_n}{n} (y\ell)^n. \tag{5.16}$$

Similarly, we can obtain

$$F_{\gamma_{B_n}}^\infty (x) \approx \frac{\varphi_n}{n} \left(\frac{x\ell}{\rho_b a_n}\right)^n. \tag{5.17}$$

Based on (5.16), we can arrive at:

$$\Phi_m^\infty \approx \frac{\varphi_m}{m} \left(\frac{x\ell}{(a_m - a_n x)\rho_b}\right)^m. \tag{5.18}$$

Based on (5.18), the asymptotic CDF of γ_{B_m} can be expressed as

$$F_{\gamma_{B_m}}^\infty (x) = U\left(x - \frac{a_m}{a_n}\right) + U\left(\frac{a_m}{a_n} - x\right) \Phi_m^\infty, \tag{5.19}$$

where Φ_m^∞ is given in (5.18).

Based on (5.11), we can replace the CDF of $F_{\gamma_{B_n}}$ by the asymptotic $F_{\gamma_{B_n}}^\infty$ of (5.17). After some manipulations, we arrive at the asymptotic SOP of the n-th user formulated by the following theorem.

Theorem 5.3 *Assuming that the LUs position obeys the PPP for the ordered channels of the LUs, the asymptotic SOP of the n-th user is given by*

$$P_n^\infty (R_n) = G_n(\rho_b)^{-D_n} + o\left(\rho_b^{-D_n}\right), \tag{5.20}$$

where we have $Q_1 = \int_0^\infty \mu_{n1} e^{-\frac{\mu_{n1}\Gamma(\delta,\mu_{n2}x)}{x^\delta}} \times \left(\frac{\mu_{n2}^\delta e^{-\mu_{n2}x}}{x} + \frac{\delta\Gamma(\delta,\mu_{n2}x)}{x^{\delta+1}}\right)$

$\left(\frac{(2^{R_n}(1+x)-1)\ell}{a_n}\right)^n dx$, $G_n = \frac{\varphi_n Q_1}{n}$, *and* $D_n = n$.

Similarly, based on (5.9), we can replace the CDF of $F_{\gamma_{B_m}}$ by the asymptotic $F_{\gamma_{B_m}}^\infty$ of (5.19). Additionally, we can formulate the asymptotic SOP of the m-th user by the following theorem.

Theorem 5.4 *Assuming that the LUs position obeys the PPP for the ordered channels of the LUs, the asymptotic SOP for the m-th user is given by*

$$P_m^\infty (R_n) = G_m(\rho_b)^{-D_m} + o\left(\rho_b^{-D_m}\right), \tag{5.21}$$

where we have $Q_2 = \int_0^{\tau_m} \mu_{m1} e^{-\frac{\mu_{m1}\Gamma(\delta,\mu_{m2}x)}{x^\delta}} \left(\frac{\mu_{m2}^\delta e^{-\mu_{m2}x}}{x} + \frac{\delta\Gamma(\delta,\mu_{m2}x)}{x^{\delta+1}}\right)$

$\left(\frac{(2^{R_m}(1+x)-1)\ell}{(a_m-a_n(2^{R_m}(1+x)-1))}\right)^m dx$, $G_m = \frac{\varphi_m Q_2}{m}$, *and* $D_m = m$.

Substituting (5.20) and (5.21) into (5.13), the asymptotic SOP for the user pair can be expressed as

$$P_{mn}^\infty = P_m^\infty + P_n^\infty - P_m^\infty P_n^\infty \approx P_m^\infty G_m(\rho_b)^{-D_m}. \tag{5.22}$$

Based on Theorems 5.4 and 5.3, and upon substituting (5.20) and (5.21) into (5.14), we arrive at the following proposition.

Proposition 5.1 *For $m < n$, the secrecy diversity order can be expressed as*

$$d_s = -\lim_{\rho_b \to \infty} \frac{\log\left(P_m^\infty + P_n^\infty - P_m^\infty P_n^\infty\right)}{\log \rho_b} = m. \tag{5.23}$$

Remark 5.1 The results of (5.23) indicate that the secrecy diversity order and the asymptotic SOP for the user pair considered are determined by the m-th user.

Remark 5.1 provides insightful guidelines for improving the SOP of the networks considered by invoking user pairing among of the M users. Since the SOP of a user pair is determined by that of the one having a poor channel, it is efficient to pair the user having the best channel and the second best channel for the sake of achieving an increased secrecy diversity order.

5.2   Enhancing Security with the Aid of Artificial Noise

In order to further improve the secrecy performance, let us now consider the employment of multiple antennas at BS for generating AN in order to degrade the Es' SNR. More particularly, the BS is equipped with N_A antennas, while all LUs and Es are equipped with a single antenna each. We mask the superposed information of NOMA by superimposing AN on Es with the aid of the BS. It is assumed that the CSI of LUs are known at BS. Since the AN is in the null space of the intended LU's channel, it will not impose any effects on LUs. However, it can significantly degrade the channel and hence the capacity of Es. More precisely, the key idea of using AN as proposed in Goel and Negi (2008) can be described as follows: an orthogonal basis of C^{N_A} is generated at BS for user κ (where $\kappa \in \{m, n\}$) as a $(N_A \times N_A)$–element precoding matrix $\mathbf{U}_\kappa = [\mathbf{u}_\kappa, \mathbf{V}_\kappa]$, where we have $\mathbf{u}_\kappa = \mathbf{h}_\kappa^\dagger / \|\mathbf{h}_\kappa\|$, and $\mathbf{V}_\kappa$ is of size $N_A \times (N_A - 1)$. Here, $\mathbf{h}_\kappa$ is denoted as the intended channel between the BS and user κ. It is noted that each column of $\mathbf{V}_\kappa$ is orthogonal to $\mathbf{u}_\kappa$. Beamforming is applied at the BS for generating AN. As such, the transmitted superposed information which is masked by AN at the BS is given by

$$\sum_{\kappa \in \{m,n\}} \sqrt{a_\kappa} \mathbf{x}_\kappa = \sum_{\kappa \in \{m,n\}} \sqrt{a_\kappa} \left(s_\kappa \mathbf{u}_\kappa + \mathbf{t}_\kappa \mathbf{V}_\kappa \right), \tag{5.24}$$

where s_κ is the information-bearing signal with a variance of σ_s^2, and $\mathbf{t}_\kappa$ is the AN. Here the $(N_A - 1)$ elements of $\mathbf{t}_\kappa$ are independent identically distributed (i.i.d.) complex Gaussian random variables with a variance of σ_a^2. As such, the overall power per transmission is $P_T = P_S + P_A$, where $P_S = \theta P_T = \sigma_s^2$ is the transmission power of the desired information-bearing signal, while $P_A = (1 - \theta) P_T = (N_A - 1) \sigma_a^2$ is the transmission power of the AN. Here θ represents the power sharing coefficients between the information-bearing signal and AN. To reduce the complexity of channel ordering in this MISO system when applying the NOMA protocol, as shown in Fig. 5.2, we divide the disc D into two regions, namely, D_1 and D_2, respectively. Here, D_1 is an internal disc with radius R_{D_1}, and the group of user n is located in this region. D_2 is an external ring spanning the radius distance from R_{D_1} to R_{D_2}, and the group of user m is located in this region. In this scenario, channel ordering is unnecessary at the BS, since in this case the path loss is the dominant channel impairment. For simplicity, we assume that user n and user m are the selected user from each group in the rest of this chapter. The cell-center user n is assumed to be capable of cancelling the interference of the cell-edge user m using SIC techniques. User n and user m are randomly selected in each region for pairing them for NOMA. The combined signal at user m is given by

Fig. 5.2 Network model for
secure NOMA transmission
using AN in multiple-antenna
scenario

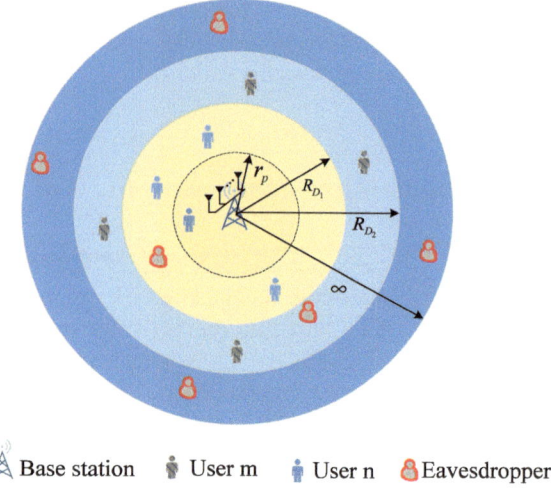

Fig. 5.2 Network model for
secure NOMA transmission
using AN in multiple-antenna
scenario

$$\mathbf{y}_m = \underbrace{\frac{\sqrt{a_m}s_m\mathbf{h}_m}{\sqrt{1+d_m^\alpha}}}_{\text{Signal part}} + \underbrace{\frac{\sqrt{a_n}s_n\mathbf{h}_m\mathbf{u}_n}{\sqrt{1+d_m^\alpha}} + \frac{\sqrt{a_n}\mathbf{h}_m\mathbf{t}_n\mathbf{V}_n}{\sqrt{1+d_m^\alpha}} + \mathbf{n}_m}_{\text{Interference and noise part}}, \tag{5.25}$$

where $\mathbf{n}_m$ is a Gaussian noise vector at user m, while d_m is the distance between
the BS and user m. Substituting (5.24) into (5.25), the received SINR at user m is
given by

$$\gamma_{B_m}^{AN} = \frac{a_m\sigma_s^2\|\mathbf{h}_m\|^2}{a_n\sigma_s^2\left|\mathbf{h}_m\frac{\mathbf{h}_n^\dagger}{\|\mathbf{h}_n\|}\right|^2 + a_n\sigma_a^2\|\mathbf{h}_m\mathbf{V}_n\|^2 + 1 + d_m^\alpha}, \tag{5.26}$$

where the variance of $\mathbf{n}_m$ is normalized to unity. As such, we can express the
transmit SNR at BS as $\rho_t = P_T$.

Since SIC is applied at user n, the interference arriving from user m can be
detected and subtracted first. The aggregate signal at user n is given by

$$\mathbf{y}_n = \underbrace{\frac{\mathbf{h}_n\sqrt{a_n}s_n}{\sqrt{1+d_n^\alpha}}}_{\text{Signal part}} + \underbrace{\frac{\mathbf{h}_n\sqrt{a_m}\mathbf{t}_m\mathbf{V}_m}{\sqrt{1+d_n^\alpha}} + \mathbf{n}_n}_{\text{Interference and noise part}}, \tag{5.27}$$

where $\mathbf{n}_n$ is the Gaussian noise at user n, while d_n is the distance between the BS
and user n. The received SINR at user n is given by

$$\gamma_{B_n}^{AN} = \frac{a_n\sigma_s^2\|\mathbf{h}_n\|^2}{a_m\sigma_a^2\|\mathbf{h}_n\mathbf{V}_m\|^2 + 1 + d_n^\alpha}, \tag{5.28}$$

where the variance of $\mathbf{n}_n$ is normalized to unity. The signal observed by Es is given by

$$\mathbf{y}_e = \sum_{\kappa \in \{m,n\}} \sqrt{a_\kappa} \mathbf{x}_\kappa \frac{\mathbf{h}_e}{\sqrt{d_e^\alpha}} + \mathbf{n}_e, \tag{5.29}$$

where $\mathbf{n}_e$ is the Gaussian noises at Es, while $\mathbf{h}_e \in \mathbb{C}^{1 \times N_A}$ is the channel vector between the BS and Es. Similar to the single-antenna scenario, again, we assume that the Es have a strong detection capability and hence they unambiguously distinguish the messages of user m and user n. The received SINR of the most detrimental E associated with detecting user κ is given by

$$\gamma_{E_\kappa}^{AN} = a_\kappa \sigma_s^2 \max_{e \in \Phi_e, d_e \geq r_p} \left\{ \frac{X_{e,\kappa}}{I_e^{AN} + d_e^\alpha} \right\}, \tag{5.30}$$

where the variance of $\mathbf{n}_e$ is normalized to unity, and we have $X_{e,\kappa} = \left| \mathbf{h}_e \frac{\mathbf{h}_\kappa^\dagger}{\|\mathbf{h}_\kappa\|} \right|^2$ as well as $I_e^{AN} = a_m \sigma_a^2 \|\mathbf{h}_e \mathbf{V}_m\|^2 + a_n \sigma_a^2 \|\mathbf{h}_e \mathbf{V}_n\|^2$.

5.2.1   New Channel Statistics

In this subsection, we derive several new channel statistics for LUs and Es in the presence of AN, which will be used for deriving the SOP in the next subsection.

Lemma 5.4 *Assuming that user m is randomly positioned in the ring D_2 of Fig. 5.2, for the case of $\theta \neq \frac{1}{N_A}$, the CDF of $F_{B_m}^{AN}$ is given by*

$$F_{B_m}^{AN}(x) = 1 - e^{-\frac{vx}{a_n}} \sum_{p=0}^{N_A-1} \frac{(vx)^p}{p!}$$

$$\times \sum_{q=0}^{p} \binom{p}{q} a_n^{q-p} a_1 \underbrace{\left(\frac{\Gamma(q+1)}{\left(vx + \frac{1}{P_S}\right)^{q+1}} - \sum_{l=0}^{N_A-2} \frac{\left(\frac{N_A-1}{P_A} - \frac{1}{P_S}\right)^l}{l!\left(vx + \frac{N_A-1}{P_A}\right)^{q+l+1}} \frac{\Gamma(q+l+1)}{} \right)}_{I(\theta)}$$

$$\times \sum_{u=0}^{p-q} \binom{p-q}{u} \frac{\gamma\left(u + \delta, \frac{vx}{a_n} R_{D_2}^\alpha\right) - \gamma\left(u + \delta, \frac{vx}{a_n} R_{D_1}^\alpha\right)}{\left(\frac{vx}{a_n}\right)^{u+\delta}}, \tag{5.31}$$

where $\gamma(\cdot, \cdot)$ is the lower incomplete Gamma function, $\Gamma(\cdot)$ is the Gamma function, $a_1 = \delta\left(1 - \frac{P_A}{(N_A-1)P_S}\right)^{1-N_A} \Big/ \left(\left(R_{D_2}^2 - R_{D_1}^2\right) P_S\right)$, and $v = \frac{a_n}{a_m P_S}$.

For the case of $\theta = \frac{1}{N_A}$, the CDF of $F_{B_m}^{AN}$ is given by (5.31) upon substituting
$I(\theta)$ by $I^*(\theta)$, where we have $I^*(\theta) = \frac{a_2 \Gamma(q+N_A)}{\left(vx+\frac{1}{P_S}\right)^{q+N_A}} \sum\limits_{u=0}^{p-q} \binom{p-q}{u}$ *and* $a_2 =$
$\frac{\delta}{\left(R_{D_2}^2 - R_{D_1}^2\right) P_S{}^{N_A}(N_A-1)!}$.

Lemma 5.5 *Assuming that user n is randomly positioned in the disc D_1 of Fig. 5.2,*
the CDF of $F_{B_n}^{AN}$ is given by

$$F_{B_n}^{AN}(x) = 1 - b_2 e^{-\frac{\vartheta x}{a_m}} \sum_{p=0}^{N_A-1} \frac{\vartheta^p x^p}{p!} \sum_{q=0}^{p} \binom{p}{q}$$

$$\times \frac{\Gamma(N_A - 1 + q)}{\left(\vartheta x + \frac{N_A-1}{P_A}\right)^{N_A-1+q} a_m^{p-q}} \sum_{u=0}^{p-q} \binom{p-q}{u} \frac{a_m^{u+\delta} \gamma\left(u + \delta, \frac{\vartheta x}{a_m} R_{D_1}^\alpha\right)}{(\vartheta x)^{u+\delta}},$$

$$(5.32)$$

where we have $b_2 = \dfrac{\delta}{R_{D_1}^2 \Gamma(N_A-1)\left(\frac{P_A}{N_A-1}\right)^{N_A-1}}$ *and* $\vartheta = \dfrac{a_m}{a_n P_S}$.

Lemma 5.6 *Assuming that the distribution of Es obeys a PPP and that the*
E-exclusion zone has a radius of r_p, the PDF of $f_{\gamma_{E_\kappa}^{AN}}$ (where $\kappa \in \{m, n\}$) is
given by

$$f_{\gamma_{E_\kappa}^{AN}}(x) = -e^{\Theta_\kappa \Psi_{\kappa 1}} \left(\frac{\left(\mu_{\kappa 2}^{AN}\right)^\delta e^{-x\mu_{\kappa 2}^{AN}}}{x} \Psi_{\kappa 1} + \frac{\delta \Theta_\kappa \Psi_{\kappa 1}}{x} + \Theta_\kappa \Psi_{\kappa 2} \right), \qquad (5.33)$$

where $\Theta_\kappa = \frac{\Gamma\left(\delta, x\mu_{\kappa 2}^{AN}\right)}{x^\delta}$, $\Gamma(\cdot, \cdot)$ *is the upper incomplete Gamma function,* $\Psi_{\kappa 1} =$
$\Omega \frac{1}{\left(\frac{x}{a_\kappa P_S}+\tau_i\right)^j}$, $\Psi_{\kappa 2} = \Omega \frac{1}{\left(\frac{x}{a_\kappa P_S}+\tau_i\right)^j} \left(\frac{j}{\left(\frac{x}{a_\kappa P_S}+\tau_i\right)} \frac{1}{a_\kappa P_S} \right)$,
$\Omega = (-1)^{N_A} \mu_{\kappa 1}^{AN} \prod\limits_{i=1}^{2} \tau_i{}^{N_A-1} \sum\limits_{i=1}^{2} \sum\limits_{j=1}^{N_A-1} a_{N_A-j,N_A-1}(2\tau_i - L)^{j-(2N_A-2)}$, $L = \tau_1 +$
$\tau_2, \tau_1 = \frac{N_A-1}{a_m P_A}, \tau_2 = \frac{N_A-1}{a_n P_A}, a_{N_A-j,N_A-1} = \binom{2N_A-j-3}{N_A-j-1}, \mu_{\kappa 1}^{AN} = \pi \lambda_e \delta(a_\kappa P_S)^\delta$, *and*
$\mu_{\kappa 2}^{AN} = \frac{r_p^\alpha}{a_\kappa P_S}$.

5.2.2 Secrecy Outage Probability

In this subsection, we investigate the SOP of a multiple-antenna-aided scenario
relying on AN. Using the results of Lemmas 5.4 and 5.6, based on (5.9), we
expressed the SOP of user m using the following theorem:

Theorem 5.5 *Assuming that the LUs and Es distribution obey PPPs and that AN is generated at the BS, for the case $\theta \neq \frac{1}{N_A}$, the SOP of user m is given by*

$$
P_m^{AN}(R_m) = \int_0^\infty -e^{\Theta_m \Psi_{m1}} \left(\frac{\left(\mu_{m2}^{AN}\right)^\delta e^{-x\mu_{m2}^{AN}}}{x} \Psi_{m1} + \frac{\delta \Theta_m \Psi_{m1}}{x} + \Theta_m \Psi_{m2} \right)
$$

$$
\times \underbrace{\left(1 - a_1^* \sum_{p=0}^{N_A-1} \frac{\iota_m^p}{p!} \sum_{q=0}^p \binom{p}{q} a_n^q \left(\frac{\Gamma(q+1)}{\left(a_n \iota_{m*} + \frac{1}{P_S}\right)^{q+1}} - \sum_{l=0}^{N_A-2} \frac{\frac{1}{l!}\left(\frac{N_A-1}{P_A} - \frac{1}{P_S}\right)^l \Gamma(q+l+1)}{\left(a_n \iota_{m*} + \frac{N_A-1}{P_A}\right)^{q+l+1}} \right) T_1^* \right)}_{K(\theta)} dx,
$$

$$(5.34)$$

where   we   have $a_1^* = \dfrac{\delta e^{-\iota_{m*}}\left(1 - \frac{P_A}{(N_A-1)P_S}\right)^{1-N_A}}{\left(R_{D_2}^2 - R_{D_1}^2\right)P_S}$, $T_1^* = \displaystyle\sum_{u=0}^{p-q} \binom{p-q}{u}$

$\dfrac{\gamma\left(u+\delta, \iota_{m*} R_{D_2}^\alpha\right) - \gamma\left(u+\delta, \iota_{m*} R_{D_1}^\alpha\right)}{\iota_{m*}^{u+\delta}}$, *and* $\iota_{m*} = \dfrac{\nu\left(2^{R_m}(1+x)-1\right)}{a_n}$.

For the case of $\theta = \frac{1}{N_A}$, *the SOP for user m is given by (5.34) upon substituting*

$K(\theta)$ *with* $K^*(\theta)$, *where* $K^*(\theta) = 1 - a_2^* \displaystyle\sum_{p=0}^{N_A-1} \frac{\iota_{m*}^p}{p!} \sum_{q=0}^p \binom{p}{q} \frac{\Gamma(q+N_A)a_n^q}{\left(a_n \iota_{m*} + \frac{1}{P_S}\right)^{q+N_A}} \sum_{u=0}^{p-q}$

$\binom{p-q}{u} T_1^*$, *and* $a_2^* = \dfrac{\delta e^{-\iota_{m*}}}{\left(R_{D_2}^2 - R_{D_1}^2\right) P_S N_A (N_A-1)!}$.

Similarly, using the results of Lemmas 5.5 and 5.6, as well as (5.11), we expressed the SOP of user n by the following theorem:

Theorem 5.6 *Assuming that the LUs and Es distribution obey PPPs and that AN is generated at the BS, the SOP of user n is given by*

$$
P_n^{AN}(R_n) = \int_0^\infty -e^{\Theta_n \Psi_{n1}} \left(\frac{\left(\mu_{n2}^{AN}\right)^\delta e^{-x\mu_{n2}^{AN}}}{x} \Psi_{n1} + \frac{\delta \Theta_n \Psi_{n1}}{x} + \Theta_n \Psi_{n2} \right)
$$

$$
\times \left(1 - b_2 e^{-\iota_n} \sum_{p=0}^{N_A-1} \frac{\iota_{n*}^p}{p!} \sum_{q=0}^p \binom{p}{q} \frac{\Gamma(N_A - 1 + q) a_m^q}{\left(a_m \iota_{n*} + \frac{N_A-1}{P_A}\right)^{N_A-1+q}} \right.
$$

$$
\left. \times \sum_{u=0}^{p-q} \binom{p-q}{u} \frac{\gamma\left(u+\delta, \iota_{n*} R_{D_1}^\alpha\right)}{\iota_{n*}^{u+\delta}} \right) dx, \qquad (5.35)
$$

where $\iota_{n*} = \dfrac{\vartheta\left(2^{R_n}(1+x)-1\right)}{a_m}$.

Based on (5.34) and (5.35), the SOP for the selected user pair can be expressed as

$$
P_{mn}^{AN} = 1 - \left(1 - P_m^{AN}\right)\left(1 - P_n^{AN}\right). \qquad (5.36)
$$

5.2.3   Large Antenna Array Analysis

In this subsection, we investigate the system's asymptotic behavior when the BS is equipped with large antenna arrays. It is noted that for the exact SOP derived in (5.34) and (5.35), as N_A increases, the number of summations in the equations will increase exponentially, which imposes an excessive complexity. Motivated by this, we seek good approximations for the SOP associated with a large N_A. With the aid of the theorem of large values, we have the following approximations: $\lim_{N_A\to\infty}\|\mathbf{h}_n\|^2 \to N_A$, $\lim_{N_A\to\infty}\|\mathbf{h}_m\|^2 \to N_A$, $\lim_{N_A\to\infty}\|\mathbf{h}_n\mathbf{V}_m\|^2 \to N_A - 1$, and $\lim_{N_A\to\infty}\|\mathbf{h}_m\mathbf{V}_n\|^2 \to N_A - 1$.

We first derive the asymptotic CDF of user n for $N_A \to \infty$. Based on (5.28), we can express the asymptotic CDF of $F_{B_n,\infty}^{AN}$ as

$$F_{B_n,\infty}^{AN}(x) = \Pr\left\{\frac{a_n P_S N_A}{a_m P_A + 1 + d_n^\alpha} \le x\right\}. \tag{5.37}$$

After some further mathematical manipulations, we can obtain the CDF of $F_{B_n,\infty}^{AN}$ for large antenna arrays in the following lemma.

Lemma 5.7 *Assuming that user n is randomly located in the disc D_1 of Fig. 5.2 and $N_A \to \infty$, the CDF of $F_{B_n,\infty}^{AN}$ is given by*

$$F_{B_n,\infty}^{AN}(x) = \begin{cases} 0, x < \zeta_n \\ 1 - \frac{\left(\frac{a_n P_S N_A}{x} - a_m P_A - 1\right)^\delta}{R_{D_1}^2}, \zeta_n \le x \le \xi_n \\ 1, x \ge \xi_n \end{cases} \tag{5.38}$$

where we have $\zeta_n = \frac{a_n P_S N_A}{R_{D_1}^\alpha + a_m P_A + 1}$ and $\xi_n = \frac{a_n P_S N_A}{a_m P_A + 1}$.

Similarly, based on (5.26), the CDF of the asymptotic $F_{B_m,\infty}^{AN}$ is given by

$$F_{B_m,\infty}^{AN}(x) = \Pr\left\{\frac{a_m P_S N_A}{a_n P_S\left|\mathbf{h}_m\frac{\mathbf{h}_n^\dagger}{\|\mathbf{h}_n\|}\right|^2 + a_n P_A + 1 + d_m^\alpha} \le x\right\}. \tag{5.39}$$

After some further mathematical manipulations, we obtain the CDF of $F_{B_m,\infty}^{AN}$ for large antenna arrays using the following lemma.

Lemma 5.8 *Assuming that user m is randomly located in the ring D_2 of Fig. 5.2 and $N_A \to \infty$, the CDF of $F_{B_m,\infty}^{AN}$ is given by*

$$
F_{B_m,\infty}^{AN}(x) = \begin{cases} 1, x \geq \zeta_{m1} \\ \dfrac{R_{D_2}^2 - t_m^2 + b_1 e^{-\frac{a_m P_S N_A}{x a_n P_S}}}{R_{D_2}^2 - R_{D_1}^2} \\ \quad \times \int_{R_{D_1}}^{t_m} r e^{\frac{r^\alpha}{a_n P_S}} dr, \zeta_{m2} < x \leq \zeta_{m1} \\ \dfrac{b_1 e^{-\frac{a_m P_S N_A}{x a_n P_S}}}{R_{D_2}^2 - R_{D_1}^2} \int_{R_{D_1}}^{R_{D_2}} r e^{\frac{r^\alpha}{a_n P_S}} dr, x < \zeta_{m2} \end{cases}, \tag{5.40}
$$

where we have $b_1 = 2e^{\frac{a_n P_A + 1}{a_n P_S}}$, $t_m = \sqrt[\alpha]{\frac{a_m P_S N_A}{x} - a_n P_A - 1}$, $\zeta_{m1} = \frac{a_m P_S N_A}{R_{D_1}^\alpha + a_n P_A + 1}$, $\zeta_{m2} = \frac{a_m P_S N_A}{R_{D_2}^\alpha + a_n P_A + 1}$, *and* $\xi_m = \frac{a_m P_S N_A}{a_n P_A + 1}$.

Let us now turn our attention to the derivation of the Es' PDF in a large-scale antenna scenario. Using the theorem of large values, we have $\lim_{N_A \to \infty} I_{e,\infty}^{AN} = a_m \sigma_a^2 \|\mathbf{h}_e \mathbf{V}_m\|^2 + a_n \sigma_a^2 \|\mathbf{h}_e \mathbf{V}_n\|^2 \to P_A$. The asymptotic CDF of $F_{\gamma_{E_K,\infty}^{AN}}$ associated with $N_A \to \infty$ is given by

$$
F_{\gamma_{E_K,\infty}^{AN}}(x) = \Pr\left\{ \max_{e \in \Phi_e, d_e \geq r_p} \left\{ \frac{a_\kappa P_S X_{e,\kappa}}{I_{e,\infty}^{AN} + d_e^\alpha} \right\} \leq x \right\}
$$

$$
= E_{\Phi_e}\left\{ \prod_{e \in \Phi_e, d_e \geq r_p} F_{X_{e,\kappa}}\left(\frac{(P_A + d_e^\alpha) x}{a_\kappa P_S} \right) \right\}. \tag{5.41}
$$

We apply the generating function and switch to polar coordinates. Then with the help of Gradshteyn and Ryzhik (2000, Eq. (3.381.9)), (5.41) can be expressed as

$$
F_{\gamma_{E_K,\infty}^{AN}}(x) = \exp\left[-\frac{\mu_{\kappa1}^{AN} \Gamma(\delta, \mu_{\kappa2}^{AN} x)}{x^\delta} e^{-\frac{P_A x}{a_\kappa P_S}} \right]. \tag{5.42}
$$

Taking derivative of (5.42), we obtain the PDF of $f_{\gamma_{E_K,\infty}^{AN}}$ in the following lemma.

Lemma 5.9 *Assuming that the Es distribution obeys a PPP and that AN is generated at the BS, the E-exclusion zone has a radius of* r_p, *and* $N_A \to \infty$, *the PDF of* $f_{\gamma_{E_K,\infty}^{AN}}$ *(where* $\kappa \in \{m, n\}$*) is given by*

$$
f_{\gamma_{E_K,\infty}^{AN}}(x) = e^{-\frac{\mu_{\kappa1}^{AN} \Gamma\left(\delta, \mu_{\kappa2}^{AN} x\right) e^{-\frac{P_A x}{a_\kappa P_S}}}{x^\delta}} - \frac{P_A x}{a_\kappa P_S} \mu_{\kappa1}^{AN} x^{-\delta}
$$

$$
\times \left(\left(\mu_{\kappa2}^{AN}\right)^\delta x^{\delta-1} e^{-\mu_{\kappa2}^{AN} x} + \Gamma\left(\delta, \mu_{\kappa2}^{AN} x\right) \left(\frac{P_A}{a_\kappa P_S} + \frac{\delta}{x} \right) \right). \tag{5.43}
$$

Remark 5.2 The results derived in (5.43) show that the PDF of $f_{\gamma_{E_K,\infty}^{AN}}$ is indepen-dent of the number of antennas N_A in our large antenna array analysis. This indicates that N_A has no effect on the channel of the Es, when the number of antennas is sufficiently high.

Let us now derive the SOP for our large antenna array scenario. Using the results of Lemmas 5.8 and 5.9, based on (5.9), we can express the SOP for user m in the following theorem.

Theorem 5.7 *Assuming that the LUs and Es distribution obey PPPs, AN is generated at the BS, and $N_A \to \infty$, the SOP for user m is given by*

$$P_{m,\infty}^{AN}(R_m) = 1 - e^{-\frac{\mu_{\kappa1}^{AN}\Gamma\left(\delta,\mu_{\kappa2}^{AN}\chi_{m1}\right)}{(\chi_{m1})^\delta} - \frac{P_A\chi_{m1}}{a_\kappa P_S}} + \frac{\mu_{m1}^{AN}b_1\Lambda_1}{R_{D_2}^2 - R_{D_1}^2}$$

$$\times \int_0^{\chi_{m2}} e^{-\frac{\mu_{m1}^{AN}\Gamma\left(\delta,\mu_{m2}^{AN}x\right)e^{-\frac{P_A x}{a_m P_S}}}{x^\delta} - \frac{a_m P_S N_A}{(2^{R_m}(1+x)-1)a_n P_S} - \frac{P_A x}{a_m P_S}} \Xi_1 dx$$

$$+ \frac{\mu_{m1}^{AN}}{R_{D_2}^2 - R_{D_1}^2} \int_{\chi_{m2}}^{\chi_{m1}} e^{-\frac{\mu_{m1}^{AN}\Gamma\left(\delta,\mu_{m2}^{AN}x\right)e^{-\frac{P_A x}{a_m P_S}}}{x^\delta} - \frac{P_A x}{a_m P_S}}$$

$$\times \left(R_{D_2}^2 - t_{m*}^2 + b_1 e^{-\frac{a_m P_S N_A}{(2^{R_m}(1+x)-1)a_n P_S}}\right) \Xi_1 \Lambda_2 dx, \qquad (5.44)$$

where we have $\Xi_1 = x^{-\delta}\left(\mu_{m2}^{AN}\left(\mu_{m2}^{AN}x\right)^{\delta-1}e^{-\mu_{m2}^{AN}x} + \Gamma\left(\delta,\mu_{m2}^{AN}x\right)\left(\frac{P_A}{a_m P_S} + \frac{\delta}{x}\right)\right)$, $\Lambda_1 = \int_{R_{D_1}}^{R_{D_2}} re^{\frac{r^\alpha}{a_n P_S}} dr$, $\Lambda_2 = \int_{R_{D_1}}^{t_{m*}} re^{\frac{r^\alpha}{a_n P_S}} dr$, $t_{m*} = \sqrt[\alpha]{\frac{a_m P_S N_A}{2^{R_m}(1+x)-1} - a_n P_A - 1}$, *and* $\chi_{m2} = \frac{\zeta_{m2}+1}{2^{R_m}} - 1$.

Similarly, using the results of Lemmas 5.7 and 5.9, as well as (5.11), we can express the SOP for user n in the following theorem.

Theorem 5.8 *Assuming that the LUs and Es distribution obey PPPs, AN is generated at the BS and $N_A \to \infty$, the SOP for user n is given by*

$$P_{n,\infty}^{AN}(R_n) = 1 - e^{-\frac{\mu_{n1}^{AN}\Gamma\left(\delta,\mu_{n2}^{AN}\chi_{n2}\right)}{(\chi_{n2})^\delta} - \frac{P_A\chi_{n2}}{a_n P_S}}$$

$$+ \mu_{n1}^{AN}\int_{\chi_{n1}}^{\chi_{n2}} e^{-\frac{\mu_{n1}^{AN}\Gamma\left(\delta,\mu_{n2}^{AN}x\right)e^{-\frac{P_A x}{a_n P_S}}}{x^\delta} - \frac{P_A x}{a_n P_S}} \Xi_2$$

$$\times \left(1 - \frac{1}{R_{D_1}^2}\left(\frac{a_n P_S N_A}{2^{R_n}(1+x)-1} - a_m P_A - 1\right)^\delta\right) dx, \qquad (5.45)$$

where $\chi_{n1} = \frac{\zeta_n+1}{2^{R_n}} - 1$, $\chi_{n2} = \frac{\xi_n+1}{2^{R_n}} - 1$, *and*

$$\Xi_2 = x^{-\delta} \left(\left(\mu_{n2}^{AN}\right)^{\delta} x^{\delta-1} e^{-\mu_{n2}^{AN} x} + \Gamma\left(\delta, \mu_{n2}^{AN} x\right) \left(\frac{P_A}{a_n P_S} + \frac{\delta}{x}\right)\right).$$

Based on (5.44) and (5.45), the SOP for the selected user pair can be expressed as

$$P_{mn,\infty}^{AN} = 1 - \left(1 - P_{m,\infty}^{AN}\right)\left(1 - P_{n,\infty}^{AN}\right). \tag{5.46}$$

5.2.4   Numerical Results

In this section, our numerical results are presented for characterizing the performance of large-scale networks. It is assumed that the power allocation coefficients of NOMA are $a_m = 0.6$ and $a_n = 0.4$. The targeted data rates of the selected NOMA user pair are assumed to be $R_m = R_n = 0.1$ bit per channel use (BPCU).

5.2.5   Secrecy Outage Probability with Channel Ordering

In Fig. 5.3, we investigate the secrecy performance in conjunction with channel ordering, which corresponds to the considered scenario considered.

Figure 5.3a plots the SOP of a single user (m-th and n-th) versus ρ_b for different user zone radii. The curves represent the exact analytical SOP of both the m-th user and n-th user derived in (5.10) and (5.12), respectively. The asymptotic analytical SOP of both the m-th and n-th users are derived in (5.21) and (5.20), respectively. Monte Carlo simulations are used for verifying our derivations. Figure 5.3a confirms the close agreement between the simulation and analytical results. A specific observation is that the reduced SOP can be achieved by reducing the radius of the user zone, since a smaller user zone leads to a lower path-loss. Another observation is that the n-th user has a more steep slope than the m-th user. This is due to the fact that we have $m < n$ and the m-th user as well as n-th user achieve a secrecy diversity order of m and n respectively, as inferred from (5.21) and (5.20).

Figure 5.3b plots the SOP of the selected user pair versus the transmit SNR ρ_b for different path-loss factors. The exact analytical SOP curves are plotted from (5.13). The asymptotic analytical SOP curves are plotted from (5.22). It can be observed that the two kinds of dashed curves have the same slopes. By contrast, the solid curves indicate a higher secrecy outage slope, which is due to the fact that the secrecy diversity order of the user pair is determined by that of the poor one. This phenomenon is also confirmed by the insights in Remark 5.1.

Figure 5.3c plots the SOP of the selected user pair versus r_p for different densities of the Es. We can observe that as expected, the SOP decreases, as the radius of the E-exclusion zone increases. Another option for enhancing the PLS is to reduce the radius of the user zone, since it reduces the total path loss. It is also worth noting

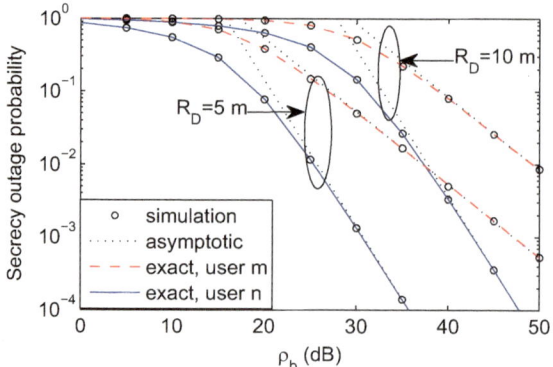

(a) The SOP versus ρ_b, with $\rho_e = 10$ dB, $\alpha = 4$, $\lambda_e = 10^{-3}$, $M = 2$, $m = 1$, $n = 2$, and $r_p = 10$ m.

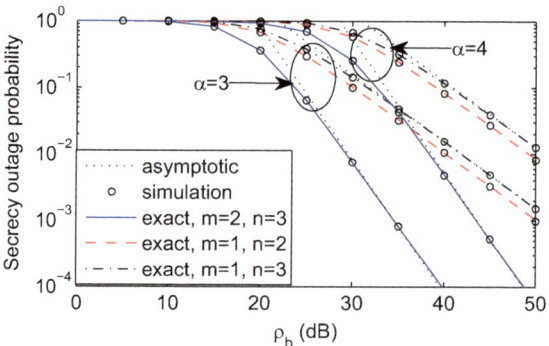

(b) The SOP of user pair versus ρ_b, with $\rho_e = 10$ dB, $\lambda_e = 10^{-3}$, $R_D = 10$ m, $M = 3$, and $r_p = 10$ m.

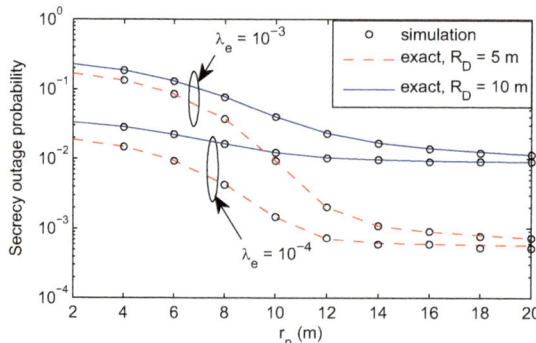

(c) The SOP of user pair versus r_p, with $\rho_b = 50$ dB, $\rho_e = 40$ dB, $M = 2$, $m = 1$, $n = 2$, and $\alpha = 4$.

Fig. 5.3 The SOP with channel ordering. (**a**) The SOP versus ρ_b, with $\rho_e = 10$ dB, $\alpha = 4$, $\lambda_e = 10^{-3}$, $M = 2$, $m = 1$, $n = 2$, and $r_p = 10$ m. (**b**) The SOP of user pair versus ρ_b, with $\rho_e = 10$ dB, $\lambda_e = 10^{-3}$, $R_D = 10$ m, $M = 3$, and $r_p = 10$ m. (**c**) The SOP of user pair versus r_p, with $\rho_b = 50$ dB, $\rho_e = 40$ dB, $M = 2$, $m = 1$, $n = 2$, and $\alpha = 4$

that having a lower E density λ_e results in an improved PLS, i.e., reduced SOP. This behavior is due to the plausible fact that a lower λ_e results in having less Es, which degrades the multiuser diversity gain, when the most detrimental E is selected. As a result, the destructive capability of the most detrimental E is reduced and hence the SOP is improved.

5.2.6 Secrecy Outage Probability with Artificial Noise

In Fig. 5.4, we investigate the secrecy performance in the presence of AN.

Figure 5.4a plots the SOP of user m and user n versus θ for different E-exclusion zones. The solid and dashed curves represent the analytical performance of user m and user n, corresponding to the results derived in (5.34) and (5.35). Monte Carlo simulations are used for verifying our derivations. Figure 5.4a confirms a close agreement between the simulation and analytical results. Again, a reduced SOP can be achieved by increasing the E-exclusion zone, which degrades the channel conditions of the Es. Another observation is that user n achieves a lower SOP than user m, which is explained as follows: (1) user n has better channel conditions than user m, owing to its lower path loss; and (2) user n is capable of cancelling the interference imposed by user m using SIC techniques, while user m suffers from the interference inflicted by user n. It is also worth noting that the SOP is not a monotonic function of θ. This phenomenon indicates that there exists an optimal value for power allocation, which depends on the system parameters.

Figure 5.4b plots the SOP of user m and user n versus λ_e for different number of antennas. We can observe that the SOP decreases, as the E density is reduced. This behavior is caused by the fact that a lower λ_e leads to having less Es, which reduces the multiuser diversity gain, when the most detrimental E is considered. As a result, the distinctive capability of the most detrimental E is reduced and hence the secrecy performance is improved. It is also worth noting that increasing the number of antennas is capable of increasing the secrecy performance. This is due to the fact that $\|\mathbf{h}_m\|^2$ in (5.26) and $\|\mathbf{h}_n\|^2$ in (5.28) both follow *Gamma* $(N_A, 1)$ distributions, which is the benefit of the improved multi-antenna diversity gain.

Figure 5.4c plots the SOP of the selected user pair versus N_A for different path loss exponents. In this figure, the curves representing the case without AN are generated by setting $\theta = 1$, which means that all the power is allocated to the desired signal. In this case, the BS only uses beamforming for transmitting the desired signals and no AN is generated. The curves in the presence of AN are generated by setting $\theta = 0.9$. We show that the PLS can be enhanced by using AN. This behavior is caused by the fact that at the receiver side, user m and user n are only affected by the AN generated by each other. By contrast, the Es are affected by the AN of both user m and user n. We can observe that the SOP of the selected user pair decreases, as the E-exclusion radius increases.

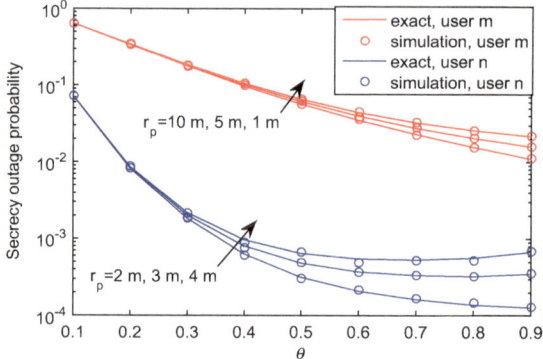

(a) The SOP versus θ, with $\alpha = 4$, $R_{D_1} = 5$ m, $R_{D_2} = 10$ m, $\lambda_e = 10^{-4}$, $N_A = 4$, $\rho_t = 30$ dB.

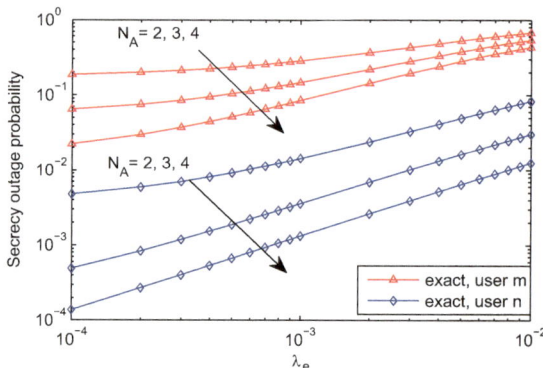

(b) The SOP versus λ_e, with $\theta = 0.8$, $\alpha = 4$, $R_{D_1} = 5$ m, $R_{D_2} = 10$ m, $\rho_t = 30$ dB, $r_p = 4$ m.

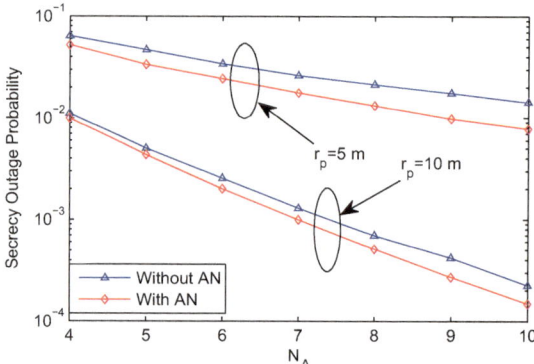

(c) The SOP of the user pair versus N_A, with $R_{D_1} = 5$ m, $R_{D_2} = 10$ m, $\alpha = 3$, $\lambda_e = 10^{-3}$, $\rho_t = 30$ dB.

Fig. 5.4 The SOP with artificial noise. (**a**) The SOP versus θ, with $\alpha = 4$, $R_{D_1} = 5$ m, $R_{D_2} = 10$ m, $\lambda_e = 10^{-4}$, $N_A = 4$, $\rho_t = 30$ dB. (**b**) The SOP versus λ_e, with $\theta = 0.8$, $\alpha = 4$, $R_{D_1} = 5$ m, $R_{D_2} = 10$ m, $\rho_t = 30$ dB, $r_p = 4$ m. (**c**) The SOP of the user pair versus N_A, with $R_{D_1} = 5$ m, $R_{D_2} = 10$ m, $\alpha = 3$, $\lambda_e = 10^{-3}$, $\rho_t = 30$ dB

5.3 Summary

This chapter discusses the enhanced security of NOMA networks.

References

Ding, Z., Liu, Y., Choi, J., Sun, Q., Elkashlan, M., Chih-Lin., I., et al. (2017). Application of non-orthogonal multiple access in LTE and 5G networks. *IEEE Communications Magazine, 55,* 185–191.

Ding, Z., Yang, Z., Fan, P., & Poor, H. V. (2014). On the performance of non-orthogonal multiple access in 5G systems with randomly deployed users. *IEEE Signal Processing Letters, 21,* 1501–1505.

Feng, Y., Yan, S., Liu, C., Yang, Z., & Yang, N. (2019). Two-stage relay selection for enhancing physical layer security in non-orthogonal multiple access. *IEEE Transactions on Information Forensics and Security, 14,* 1670–1683.

Goel, S., & Negi, R. (2008). Guaranteeing secrecy using artificial noise. *IEEE Transactions on Wireless Communications, 7,* 2180–2189.

Gradshteyn, I. S., & Ryzhik, I. M. (2000). *Table of integrals, series and products* (6th ed.). New York: Academic Press.

He, B., Liu, A., Yang, N., & Lau, V. K. (2017). On the design of secure non-orthogonal multiple access systems. *IEEE Journal on Selected Areas in Communications, 35,* 2196–2206.

Liu, Y., Wang, L., Zaidi, S., Elkashlan, M., & Duong, T. (2016). Secure D2D communication in large-scale cognitive cellular networks: A wireless power transfer model. *IEEE Transactions on Communications, 64,* 329–342.

Mukherjee, A., Swindlehurst, A. (2011). Robust beamforming for security in MIMO wiretap channels with imperfect CSI. *IEEE Transactions on Signal Processing, 59,* 351–361.

Qin, Z., Gao, Y., & Plumbley, M. D. (2018). Malicious user detection based on low-rank matrix completion in wideband spectrum sensing. *IEEE Transactions on Signal Processing, 66,* 5–17.

Saito, Y., Kishiyama, Y., Benjebbour, A., Nakamura, T., Li, A., & Higuchi, K. (2013). Non-orthogonal multiple access (NOMA) for cellular future radio access. In IEEE Proceeding of Vehicular Technology Conference (VTC), Dresden, Germany (pp. 1–5).

Tekin, E., & Yener, A. (2008). The general Gaussian multiple-access and two-way wiretap channels: Achievable rates and cooperative jamming. *IEEE Transactions on Information Theory, 54,* 2735–2751.

Wyner, A. D. (1975). The wire-tap channel. *Bell System Technical Journal, 54,* 1355–1387.

Zhang, X., Zhou, X., & McKay, M. R. (2013). Enhancing secrecy with multi-antenna transmission in wireless ad hoc networks. *IEEE Transactions on Information Forensics and Security, 8,* 1802–1814.

Zhou, X., & McKay, M. R. (2010). Secure transmission with artificial noise over fading channels: Achievable rate and optimal power allocation. *IEEE Transactions on Vehicular Technology, 59,* 3831–3842.

Chapter 6
Artificial Intelligence (AI) Enabled NOMA

Recently, machine learning has been extensively applied in various areas including wireless communications. The more recent work has shown the power of deep learning in physical layer communications (Qin et al. 2019) and resource allocation (Ye et al. 2018). In this chapter, we will discuss the adaptive NOMA enabled by artificial intelligence AI and the new application of NOMA in the unmanned aerial vehicle (UAV) networks, with the goal to provide a potential solution to realize UAV networks with NOMA.

6.1 AI for Adaptive NOMA

6.1.1 Unified NOMA

In this part, we propose a unified NOMA framework, which contains both PD-NOMA and CD-NOMA techniques. As shown in Fig. 6.1, we first map the superposed signals of multiple users to single RB or multi-RB over a sparse matrix, in which most of the elements are zero. Note that single RB and multi-RB correspond to single carrier and multi-carrier, respectively. In other words, single carrier NOMA (PD-NOMA) is the special case of multi-carrier NOMA (CD-NOMA). The rows and columns of the sparse matrix represent different RB and different users, respectively. In the matrix of Fig. 6.1, "1" represents the user occupies the corresponding RB and "0" otherwise. The use of such a sparse matrix is essential to capture the features of SCMA (Nikopour and Baligh 2013) and pattern division multiple access (PDMA) (Chen et al. 2016), where the optimal design of sparse matrix for CD-NOMA is capable of reducing detection complexity at receivers. In particular, SCMA and PDMA belong to CD-NOMA's different types of forms, in which the equal and unequal column weight sparse matrixes are employed, respectively.

Y. Liu et al., *Non-Orthogonal Multiple Access for Massive Connectivity*,
SpringerBriefs in Computer Science, https://doi.org/10.1007/978-3-030-30975-6_6

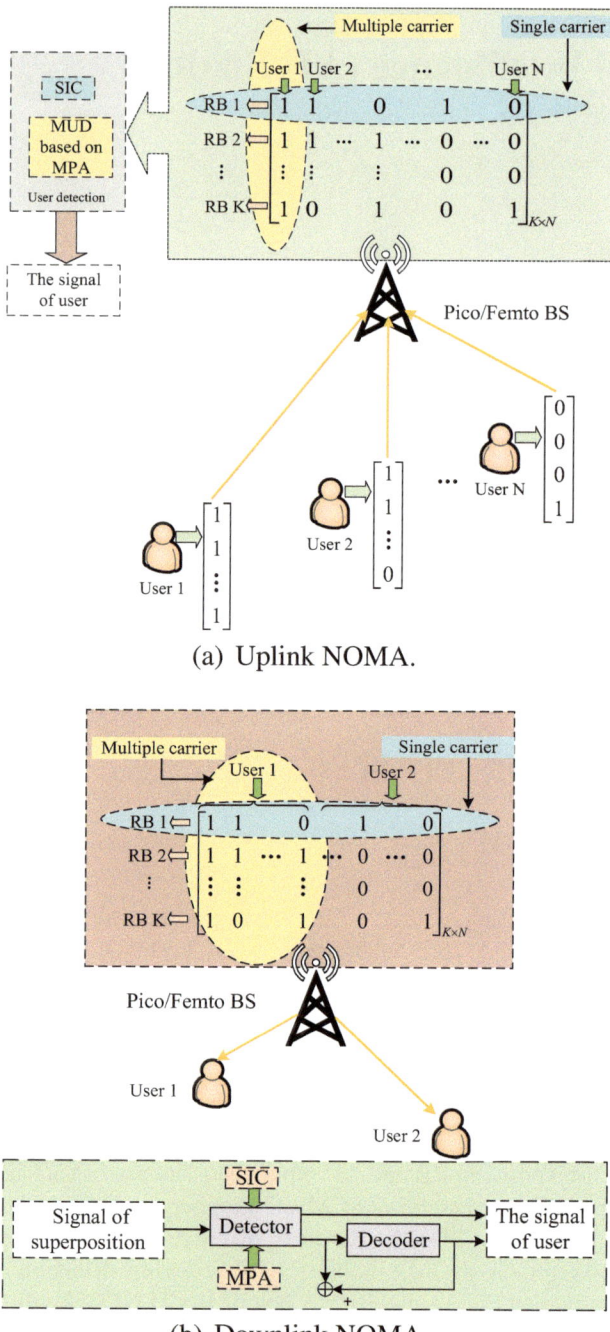

(a) Uplink NOMA.

(b) Downlink NOMA.

Fig. 6.1 (**a**) Uplink and (**b**) downlink NOMA systems

Regarding PD-NOMA, a transmitter is capable of multiplexing multiple users via different power levels within the single subcarrier, i.e., the first row of sparse matrices illustrated in Fig. 6.1. At receivers, PD-NOMA exploits SIC to remove the multi-user interference. Additionally, PD-NOMA can be also realized in multi-carriers, with the aid of appropriate user scheduling and power allocation approaches. It is worth noting that downlink multi-user superposition transmission (MUST), which is essentially a special case of PD-NOMA, has been standardized.

CD-NOMA can be regarded as a special extension that directly maps data streams of multiple users into multiple carriers by using the sparse matrix or low density spreading code at transmitters, as shown in the multi-carrier case in Fig. 6.1. At receivers, multiple users are distinguished by MPA to obtain the multiplexing or coding gains. For example, SCMA utilizes a sparse matrix, in which each column can be selected from the predefined codebooks. With the multidimensional constellations to optimize codebooks, SCMA is able to achieve enhanced shaping and coding gains. While for PDMA, the core concept is to jointly optimize transmitters with sparse pattern design and receivers with MPA-based detection. The design of sparse pattern can provide disparate diversity for multiple users and further reduce the complexity of detection. Additionally, phase shifting is an effective way to obtain constellation shaping gain. It is worth noting that the pivotal difference between these two schemes is that the number of RBs occupied by each user has to be the same in SCMA, while PDMA allows a variable number of RBs to be occupied by the same user. For multi-user shared access (MUSA) (Yuan et al. 2016), each user's data symbols are spread by a special spread sequence to facilitate SIC implementation. This kind of spread sequence can be selected from the sparse matrix as illustrated in Fig. 6.1, which requires special design to achieve low cross-correlation. Note that multiple spreading sequences constitute a pool, where each user can select one sequence from it.

6.2 NOMA in UAV Networks

With the rapid development of control technology and manufacture business, unmanned aerial vehicles (UAVs), which were originally sparked by the military use, have gradually demonstrated the civil potentials to new applications and markets opportunities, such as advanced cargo distribution, aerial photography, and wildfire management, to name a few. UAV-aided communications have been recognized as an emerging technique on both industry and academia for its superior flexibility and autonomy. On one hand, industry projects, such as Google Loon project, Internet-delivery drone in Facebook, and airborne LTE services in AT&T, have been deployed for providing airborne global massive connectivity. On the other hand, promising research scenarios for assisting 5G and beyond can be as follows: establishing temporal communication infrastructure during natural disasters, offloading traffic for dense networks, data collection for supporting IoT networks, etc. (Liu et al. 2019).

Before introducing the UAV networks with NOMA, we characterize the unique features of UAV networks first. Generally, UAV networks have the following characteristics:

- **Path loss**: Since there are usually not many obstacles in the air, we use a simplified model to assume that the line-of-sight (LOS) links between the UAVs and the users are dominated, which are significantly less affected by shadowing and fading. In more complicated practical scenarios, such as urban areas where buildings and other obstacles on the ground may block UAV flight and signal transmission, both LOS and non-line-of-sight (NLOS) links require to be considered.
- **Mobility**: When a UAV flies around, the coverage areas become various. Therefore, the UAV can support different ground users. For example, UAVs are capable of roaming above a group of users to enhance the channel conditions so as to provide high throughput.
- **Agility**: Based on the real-time requirements from the users, UAVs can be deployed quickly and their positions can be adjusted within a 3D space flexibly, which enables UAV networks to provide flexible and on-demand service to the ground users with lower costs compared to the terrestrial BS.

To support massive connectivity in UAV networks, power-domain NOMA can be adopted to support different users over the same time/frequency slot based on the aforementioned three features. Note that UAVs communicate with ground users mainly relying on LoS links. Such characteristics make that there are no distinct channel gain differences between UAVs to multiple NOMA ground users compared to conventional terrestrial NOMA communications. Therefore, the NOMA-aided UAV networks should be carefully redesigned according to large-scale fading differences of NOMA users by invoking user pairing/grouping techniques. Figure 6.2 illustrates a few scenarios in the downlink of NOMA-aided UAV networks, spanning from the mathematical modeling, the joint resource and trajectory optimization to aerial BS placement and movement design.

Specifically, by adopting the NOMA technique in the downlink transmission, different users can share the same resource block by using different transmit power levels with a total power constraint. Taking the case shown in the middle of Fig. 6.2 as an example, where there is a UAV with two covered users, i.e., $U1$ and $U2$, the UAV sends a superimposed signal containing the signals for the two users. In order to guarantee the fairness, less power is allocated to $U1$ that is with better downlink channel state information (CSI). SIC is used for signal detection at the receiver. $U2$, with higher transmit power and poorer channel gain, is decoded first by treating $U1$ as interference. Once $U2$ is detected and decoded, its signal component will be subtracted from the received signal to facilitate the detection of $U1$. By doing so, $U2$ suffers from the higher inter-user interference and the detection error of $U2$ will pass to $U1$. Therefore, we have to allocate sufficient power to the user that is detected first, i.e., $U2$.

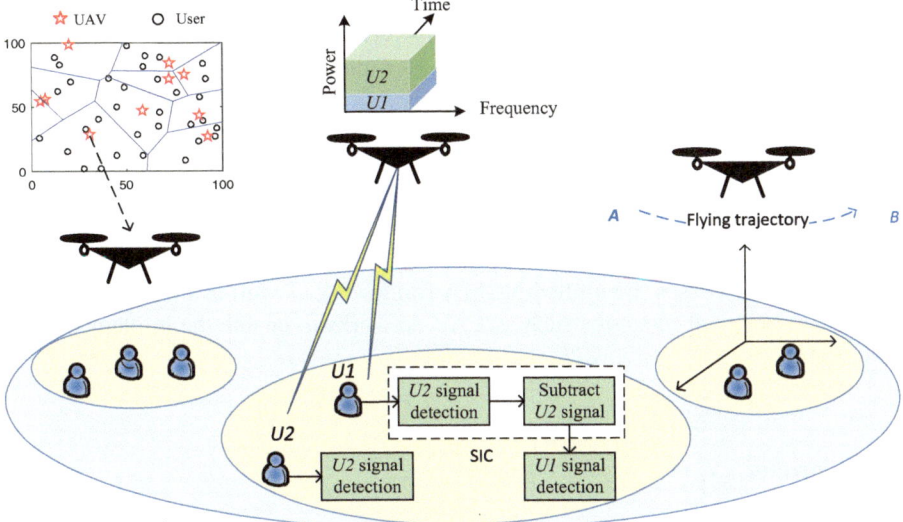

Fig. 6.2 Illustration for NOMA-aided UAV networks

The logic behind Fig. 6.2 is to provide a comprehensive understanding of the design of practical NOMA-aided UAV networks. On the left side of Fig. 6.2, the Voronoi figure shows the spatial relationship of UAVs and NOMA users, where the five-pointed stars representing the UAVs are the vertical mapping from the air to the ground. The right side of Fig. 6.2 shows the trajectory of a UAV when it serves the ground users. Figure 6.2 as a whole presents a complicated system with multiple UAVs and dynamic users. Based on the three characteristics and new requirements of various scenarios in NOMA-aided UAV networks, the following issues should be addressed accordingly:

- **Random spatial modeling:** To quantify NOMA-aided UAV networks with LOS and Nakagami-m fading channels, performance analysis of NOMA-aided networks is desired before designing and implementing such networks. Scholastic geometry is an effective tool to analyze the average performance of the networks. Using a stochastic geometry-based model, the ground users are grouped together to share the same resource with NOMA and associated with the most suitable UAV.
- **Resource allocation and trajectory design**: By considering the mobility of UAVs, trajectory design is necessary to provide better coverage for various ground users. As shown on the right side of Fig. 6.2, from the starting point, *A*, to the destination, *B*, the power allocation of the NOMA users should be addressed properly to further improve the system performance (Cui et al. 2019).
- **Dynamic movement and deployment design:** When there are multiple UAVs in the networks and the NOMA users are roaming dynamically, how to adjust the positions of UAVs dynamically to optimize the system performance with

considering dynamic user grouping of NOMA users becomes more challenging, especially when there are both the LOS and NLOS links between transmitters and receivers. The movement and deployment of multiple UAVs should be carefully designed in the dynamic environment.

6.3   Summary

This chapter discusses the unified NOMA framework as well as the application of NOMA in practical scenarios such as UAV networks to enable the implementation of NOMA.

References

Chen, S., Ren, B., Gao, Q., Kang, S., Sun, S., & Niu, K. (2016). Pattern division multiple access PDMA—A novel non-orthogonal multiple access for 5G radio networks. *IEEE Transactions on Vehicular Technology, 66*, 3185–3196.

Cui, F., Cai, Y., Qin, Z., Zhao, M., & Li, G. Y. (2019). Multiple access for mobile-UAV enabled networks: Joint trajectory design and resource allocation. *IEEE Transactions on Communications, 67*, 4980–4994.

Liu, Y., Qin, Z., Cai, Y., Gao, Y., Li, G. Y., & Nallanathan, A. (2019). UAV communications based on non-orthogonal multiple access. *IEEE Wireless Communications, 26*, 52–57.

Nikopour, H., & Baligh, H. (2013). Sparse code multiple access. In *2013 IEEE 24th Annual International Symposium on Personal, Indoor, and Mobile Radio Communications (PIMRC)* (pp. 332–336).

Qin, Z., Ye, H., Li, G. Y., & Juang, B. F. (2019). Deep learning in physical layer communications. *IEEE Wireless Communications, 26*, 93–99.

Ye, H., Liang, L., Li, G. Y., Kim, J., Lu, L., & Wu, M. (2018). Machine learning for vehicular networks: Recent advances and application examples. *IEEE Vehicular Technology Magazine, 13*, 94–101.

Yuan, Z., Yu, G., Li, W., Yuan, Y., Wang, X., & Xu, J. (2016). Multi-user shared access for internet of things. In *IEEE Proceedings of Vehicular Technology Conference (VTC)*.

Part III
Challenges and Conclusions

Chapter 7
Challenges and Conclusions

7.1 Research Challenges

There still exist some open research issues in context of the implementation of NOMA enabled UAV networks, which are highlighted as follows:

- **A unified spatial model for NOMA-aided UAV networks**: there are various communication scenarios for NOMA-aided UAV networks, e.g., single-UAV case, multiple-UAV case, uplink, downlink, cooperative communications scenarios, etc. A unified spatial analytical framework for NOMA-aided UAV networks is desired, which can be effortlessly switched to fit different practical application scenarios.

- **Data driven NOMA-aided UAV networks design**: Big data has been recognized as a powerful tool to provide insightful guidelines for real systems. Most of the current research contributions in context of NOMA-aided UAV networks are based on data generated randomly, which may be different from practical situations. Data from social networks, such as Twitter and Facebook, can be used for collecting position information of mobile users. As a further advance, more sophisticated big data analytical approaches, such as data mining and stochastic modeling, can be invoked for analyzing the historical data and providing more accurate prediction in terms of NOMA users' mobilities. By doing so, the UAVs are able to adjust their positions more accurately to further improve the system performance.

- **MIMO-NOMA design in UAV networks**: NOMA is expected to coexist with MIMO techniques for further improving the spectral efficiency and supporting massive connectivity of UAV networks. Nevertheless, applying multiple-antenna techniques in NOMA requires carefully designing the channel ordering. In contrast to the single antenna NOMA case in which the channels of users are scalars, the channels in multiple-antenna-aided NOMA are vectors or matrix.

© The Author(s), under exclusive license to Springer Nature Switzerland AG 2020
Y. Liu et al., *Non-Orthogonal Multiple Access for Massive Connectivity*,
SpringerBriefs in Computer Science, https://doi.org/10.1007/978-3-030-30975-6_7

Additionally, due to the 3D characteristics of UAV networks, beamforming-based or cluster-based MIMO-NOMA design becomes more challenging. As a result, how to order channels in MIMO-NOMA systems with considering the characteristics of UAV networks requires more research contributions.

- **Low latency design for NOMA-aided UAV networks**: It is worth pointing out that one remarkable feature of UAVs is agility. The SIC decoding characteristics of NOMA inevitably brings considerable delays at receivers if the number of NOMA users is large. A possible solution is to adopt hybrid multiple access by dividing a large number of NOMA users into different orthogonal groups. In each group, a small number of users invoke NOMA for decreasing the delay caused by SIC.

7.2   Conclusions

This book aims to discuss the recent advanced multiple-access techniques, named NOMA, from the three most important perspectives, including compatibility, sustainability, and security with the purpose of supporting massive number of devices in the future networks. Specifically, the performance analysis of the scenarios in the three aspects has been provided as an example. Moreover, the most recent application of NOMA in the UAV networks has been discussed as an example of the reality of NOMA. Finally, the potential research directions have been identified, especially in the UAV scenarios. With all the provided examples, we can obtain the insights that NOMA is one of the promising solutions to offer more access opportunities with the limited spectrum by exploiting one extra domain. With the low complexity design at the receiver, we can predict that NOMA will be a promising solution to support massive connectivity in the future wireless networks.

Index

© The Author(s), under exclusive license to Springer Nature Switzerland AG 2020
Y. Liu et al., *Non-Orthogonal Multiple Access for Massive Connectivity*,
SpringerBriefs in Computer Science, https://doi.org/10.1007/978-3-030-30975-6